ELECTROMECHANICS AND ELECTRICAL MACHINERY

J. F. Lindsay and M. H. Rashid

Department of Electrical Engineering
Concordia University, Montreal

Prentice-Hall, Inc., Englewood Cliffs, N.J. 07632

Library of Congress Cataloging in Publication Data

Lindsay, J. F.
 Electromechanics and electrical machinery.

 Bibliography: p. 225
 Includes index.
 1. Electromechanical devices. 2. Electric machinery. I. Rashid, M. H. II. Title.
TK2000.L56 1985   621.31′042   85–3672
ISBN 0–13–250093–0

Editorial/production supervision and interior design: Fred Dahl
Cover design: 20/20 Services, Inc.
Manufacturing buyer: Gordon Osbourne

Printed in the United States of America

10 9 8 7 6 5 4 3 2 1

ISBN 0-13-250093-0   01

Prentice-Hall International (UK) Limited, *London*
Prentice-Hall of Australia Pty. Limited, *Sydney*
Prentice-Hall Canada Inc., *Toronto*
Prentice-Hall Hispanoamericana, S.A., *Mexico*
Prentice-Hall of India Private Limited, *New Delhi*
Prentice-Hall of Japan, Inc., *Tokyo*
Prentice-Hall of Southeast Asia Pte. Ltd., *Singapore*
Editora Prentice-Hall do Brasil, Ltda., *Rio de Janeiro*
Whitehall Books Limited, *Wellington, New Zealand*

CONTENTS

4

ELECTROMECHANICAL ENERGY CONVERSION 56

5

TRANSFORMERS 71

6

DC MACHINES 97

PREFACE

In today's typical undergraduate engineering program, the time available for the study of electromechanical energy conversion seems to come under constant pressure. In many cases, the basic ideas must be presented to students within a single semester. Indeed, even the model curriculum produced by a committee of the Power Engineering Society of the Institute of Electrical and Electronics Engineers* allocates only one semester to this subject. This book is a response to the need for a text that is suitable for such a course.

The task of introducing such a wide range of material in a one-semester course is formidable. Yet the area offers many interesting challenges. We have therefore developed the basic models of the most common machines at a level appropriate to the junior year, at the same time pointing out the limitations of these simple models in dealing with modern adjustable speed drives. An unusual feature of the book is the final chapter, which can provide an approach to some dynamic problems as well as one steady state problem that is very cumbersome to solve using conventional models. Readers who are familiar with analogous circuits based on through and across variables will find this treatment quite straightforward.

Other than this one topic, the remainder of the book assumes a knowledge of basic electric circuits, electricity and magnetism, and mechanics.

*IEEE Task Force Report: "A model undergraduate electric power engineering curriculum," *IEEE Transactions on Power Apparatus and Systems,* Vol PAS-100, No. 6, June 1981, pp. 3110–3115.

The only other point where the treatment may not be considered quite conventional relates to problems involving nonlinear characteristics such as magnetization characteristics. The solutions given in the examples use linear interpolation to determine particular values. They are therefore appropriate for use with a computer. While it is not necessary to use a computer for such problems, the methods given may be used either for a graphical inspection of the characteristic or for interpolation within a program.

It is not possible to have a realistic study of motors that form part of an adjustable speed drive system in a course lasting one semester. However, the basic strategy of controlling the motor has been given. In addition, an indication of the basic circuits within the power converters has been presented. This will not produce experts, but it should familiarize readers with a stimulating and challenging area.

J. F. LINDSAY
M. H. RASHID
Montreal

1

INTRODUCTION

The subjects of electromechanics and electrical machinery deal with the conversion of energy between electrical and mechanical systems. For rotational motion, the devices are therefore motors or generators; for translational motion, the devices are varied and may include electromagnets for lifting ferromagnetic materials, solenoids for acting on some mechanical part, or the chimes which you may have at home as a doorbell.

Industrial installations such as the one shown in Fig. 1-1 may have both motors and generators. This is a view of a synchronous motor driving four dc generators in a steel mill. In addition to the rotating machines, the protective circuit breakers can also be seen; these usually involve the operation of a small electromagnet. Although many motors and generators such as those in Fig. 1-1 are intended for use in locations where their shafts are in a horizontal position, there are occasions where the shaft must be mounted vertically. Large hydro generators are often mounted in this way, and Fig. 1-2 shows an example of a vertical fan-cooled pump motor. In small sizes, motors are expected to operate no matter which way their shafts are directed; portable power tools are well-known examples of this requirement. There are examples of very small motors (micromotors) used to drive devices such as cameras. At all power levels there is concern over the efficiency of the conversion of energy as well as the basic mechanical characteristics. In order to predict the performance of any of these devices in a given situation, it is necessary to have a suitable model; it is the derivation and application of some of these basic models which forms the material of this text.

Figure 1-1 Synchronous motor driving four DC generators in a steel mill (Photo courtesy of Canadian General Electric Co.).

In any electromechanical device there is a need to model the electric circuit aspect, the mechanical system aspect, and the interface between them. The modeling problem is thus rather more complex than those found in previous courses. A knowledge of electric circuit analysis and basic mechanics is required. Some devices having a complicated mechanical structure turn out to have a simple circuit model; this is often the case in

Figure 1-2 Vertical fan-cooled squirrel cage induction motor (Photo courtesy of Canadian General Electric Co.).

rotating machines. Other devices which are very simple in their structure require models involving nonlinear differential equations. The scope of this text will be limited to situations where the equations are linear. The non-linearity in electromechanics is due both to the fact that some variables are not directly proportional to others and to the fact that there are products of the variables in some equations.

Virtually all electromechanical devices in use today are based on the magnetic field. There are some that are based on the electrical field, but since these have not proved amenable to use at large power levels, they are not at all common.

Since all systems involving the transmission of power use three-phase circuits, we review these first. Because magnetic circuits provide the basis of all the devices under study, we review them in Chapter 3; the application to the general principles of electromechanical energy conversion are considered in Chapter 4. Although a power transformer does not involve electromechanical energy conversion, it is convenient to study this device next, since most machines have some transformer action as part of their operation. This is done in Chapter 5.

The next three chapters consider the modeling and applications of the three most common industrial motors and generators. The first of these is the dc machine, which until recently was used in virtually all variable-speed drives. The induction machine is the least expensive and is finding increasing use in variable-speed drives. The synchronous machine is invariably used for very large generators and often for large motors.

To complete the text, Chapter 9 deals with the modeling of motors from the mechanical port (shaft). These models have been developed using methods of deriving analogous circuits based on through and across variables. For applications where it is necessary to investigate the dynamic and steady-state response resulting from changes in mechanical loading of a motor, these models can be of considerable value.

The text is not an introduction to the design of electric motors and generators. Rather, it provides a basis for designs involving the application of such machines. The models that are used require some understanding of the basic operation of each machine; without this understanding, it is not possible to assess the likely errors in the predicted performance. The points at which the basic models require refinement have been acknowledged and the need for further study emphasized. In presenting the material, SI units have been used. However, in a few problems the power has been expressed in horsepower, since this unit is still to be found. One horsepower is equivalent to 746 W.

2

THREE-PHASE CIRCUITS

2.1 INTRODUCTION

A three-phase circuit is one in which the source consists of three sinusoidal sources. When the source is *balanced,* the three sources have equal magnitudes and the phase angles between the individual sources are all 120°. This is shown in Fig. 2-1.

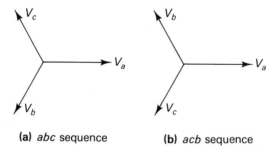

(a) *abc* sequence   **(b)** *acb* sequence

Figure 2-1 Phasor diagrams of balanced voltages.

Analysis of three-phase circuits is strictly the analysis of an ordinary circuit in which the sources take a particular form. However, the analysis of a balanced circuit can be simplified so that only one of the three phases need be considered. For this reason, a brief review of single-phase analysis is given.

2.2 SINGLE-PHASE CIRCUITS

Consider first the single-phase circuit shown in Fig. 2-2. For analysis of the sinusoidal steady state, the complex impedance $\hat{Z}$ is given by

$$\hat{Z} = R + jX = |Z| \underline{/\beta} \qquad (2\text{-}1)$$

where

$$|Z| = \sqrt{R^2 + X^2}$$

and

$$\beta = \tan^{-1} \frac{X}{R} \qquad (2\text{-}2)$$

The current (in complex form) is then given by

$$\hat{I} = \frac{|V| \underline{/\alpha}}{|Z| \underline{/\beta}} = \frac{|V|}{|Z|} \underline{/\alpha - \beta} \qquad (2\text{-}3)$$

Note. $|V|$ and $|I|$ are the magnitudes of the voltage $\hat{V}$ and the current $\hat{I}$. From this point on, the symbols V and I will be used rather than $|V|$ and $|I|$, this being the normal practice.

In this context, currents and voltages are normally given using rms values. The average rate at which power is dissipated is called the *power* (or *true power*), and it is given by

$$P = VI \cos \beta \qquad (2\text{-}4)$$

where β is the phase angle between $\hat{V}$ and $\hat{I}$.

It is also possible to define *complex power*, which is the product of the complex voltage and the *conjugate* of the complex current:

$$\hat{S} = \hat{V} \hat{I}^* \qquad (2\text{-}5)$$

$$= P + jQ \qquad (2\text{-}6)$$

where P is the power as defined above and Q is the *reactive power*.

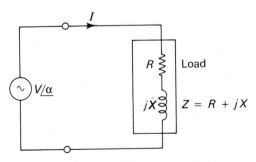

Figure 2-2 Single-phase circuit.

The reactive power (the unit is the var or kvar or Mvar) is given by

$$Q = VI \sin \beta \qquad (2\text{-}7)$$

Note:

1. The product of current and voltage (magnitudes) does *not* give power; for this reason it is always called the *apparent power* (the unit is the VA or kVA or MVA).
2. The ratio of power to apparent power is called the *power factor* (PF).
3. The ratio of reactive power to the apparent power is called the *reactive factor*.
4. The reactive power is positive for flow *into* an inductive load.
5. With an inductive load the current lags the voltage, and any numerical value for power factor must include the fact that it is inductive (or lagging).
6. With a capacitive load the current leads the voltage, and power factor is then capacitive (or leading).

Phasor diagrams for these situations are shown in Fig. 2-3.

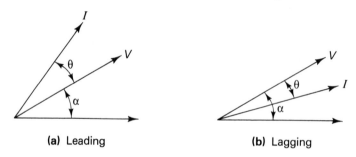

(a) Leading   (b) Lagging

Figure 2-3 Phasor diagrams.

2.3 BALANCED THREE-PHASE CIRCUITS

There are two possible ways in practice by which three individual sources can be connected to produce a three-phase source and by which load elements can be connected to produce a three-phase load. The wye (or Y) connection is shown in Fig. 2-4a; the delta connection is shown in Fig. 2-5a. In each connection it is common to consider that there are two sets of voltages and currents. One set consists of the currents and voltages of the individual phases; these are called the phase currents and voltages. The other set consists of the currents and voltages that are external to the three-phase source or load. These are the line currents and voltages. Sometimes, for clarity, the line voltages are called the line-to-line voltages, because it is the voltage between pairs of lines that is being referred to.

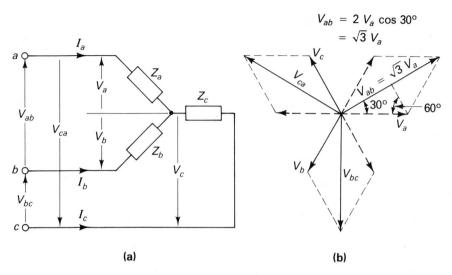

Figure 2-4 Wye-connected load.

For a wye connection, it can be seen from the circuit diagram that the currents flowing through the external lines are the same as those flowing through the individual phases, but the line voltages are not the same as the phase voltages. For a delta connection, it is the line and phase voltages that are the same, but the line and phase currents must be different.

Let the phase voltages in a wye-connected generator be represented by

$$\hat{V}_a = V \angle 0 \tag{2-8}$$

$$\hat{V}_b = V \angle -120 \tag{2-9}$$

$$\hat{V}_c = V \angle -240 = V \angle +120 \tag{2-10}$$

where the angles are given in degrees for convenience although, strictly, they should be expressed in radians. These angles are appropriate for the sequence *abc* (positive phase sequence). If the phase sequence is reversed, that is, *acb* (negative phase sequence), the angles used to describe $\hat{V}_b$ and $\hat{V}_c$ are interchanged.

If we recall that any voltage is the difference in potential between two points, it is then clear that the line voltages are the differences in voltage to datum between the different pairs of lines. From the phasor diagram of the wye connection shown in Fig. 2-4b it can be seen that the line voltages form a three-phase set in which the magnitude of each voltage is $\sqrt{3}$ times the phase voltage, and each is leading the corresponding phase voltage by

30°. Expressed algebraically, the three line voltages are given by

$$\hat{V}_{ab} = \hat{V}_a - \hat{V}_b \tag{2-11}$$

$$\hat{V}_{bc} = \hat{V}_b - \hat{V}_c \tag{2-12}$$

$$\hat{V}_{ca} = \hat{V}_c - \hat{V}_a \tag{2-13}$$

The currents flowing into a delta connection are shown in Fig. 2-5a. Applying the current law at each vertex of the delta gives the relationship between the line and phase currents, shown in Fig. 2-5b. Again, there is the factor of $\sqrt{3}$, but this time it is applied to the line currents whose magnitudes are $\sqrt{3}$ times the phase currents and lagging them by 30°.

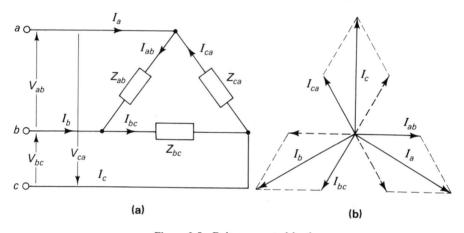

(a)   (b)

Figure 2-5   Delta-connected load.

The relevant equations are:

$$\hat{I}_a = \hat{I}_{ab} - \hat{I}_{ca} \tag{2-14}$$

$$\hat{I}_b = \hat{I}_{bc} - \hat{I}_{ab} \tag{2-15}$$

$$\hat{I}_c = \hat{I}_{ca} - \hat{I}_{bc} \tag{2-16}$$

In all these equations the order of the subscripts is significant. The current $\hat{I}_{ab}$, for example, is the current flowing through the branch from vertex a to vertex b, and is given by the normal relation:

$$\hat{I}_{ab} = \frac{\hat{V}_{ab}}{\hat{Z}_{ab}} \tag{2-17}$$

Example 2-1

A balanced three-phase load is connected to a balanced three-phase source as shown in Fig. 2-6. Both have wye connections, and the neutral conductor is connected between the neutral points of the source and the load. Obtain expressions for the line and neutral currents.

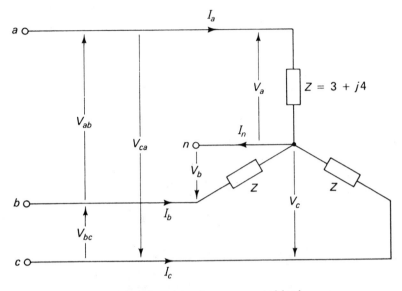

Figure 2-6 Balanced wye-connected load.

Solution.   If $\hat{V}_a$ is taken as reference, the phase voltages are defined as in Eqs. (2-8) to (2-10), and the three line (and phase) currents are:

$$\hat{I}_a = \frac{\hat{V}_a}{\hat{Z}} = \frac{V\;\underline{/0}}{Z\;\underline{/\beta}}$$

$$\hat{I}_b = \frac{\hat{V}_b}{\hat{Z}} = \frac{V\;\underline{/-120}}{Z\;\underline{/\beta}}$$

$$\hat{I}_c = \frac{\hat{V}_c}{\hat{Z}} = \frac{V\;\underline{/+120}}{Z\;\underline{/\beta}}$$

$$\hat{I}_n = \hat{I}_a + \hat{I}_b + \hat{I}_c$$

$$= \frac{V}{Z}\left[1\;\underline{/-\beta} + 1\;\underline{/-\beta-120} + 1\;\underline{/-\beta+120}\right]$$

$$= 0$$

That is, the current in the neutral connection to a balanced wye-connected load from a balanced wye-connected source is zero. The neutral connection is therefore unnecessary.

At this point we may note that whether or not the neutral is present, this problem can be solved by considering only one of the three phases, normally the reference phase. The current is given by $\hat{I} = \hat{V}/\hat{Z} = I\;\underline{/\beta}$, and the currents in the other two phases have the same magnitude, but lagging by 120° and 240°, respectively.

If the impedance, $\hat{Z}$, is $(3 + j4)$ Ω per phase and the source voltage is 550 V, the numerical values are:

$$V_{ph} = \frac{550}{\sqrt{3}} = 317.5 \text{ V}$$

so that

$$\hat{V}_a = 317.5 \; \underline{/0} \cdot \text{V}$$

and since

$$\hat{Z} = 3 + j4 = 5 \; \underline{/53.1} \; \Omega$$

$$\hat{I}_a = \frac{317.5 \; \underline{/0}}{5 \; \underline{/53.1}} = 63.5 \; \underline{/-53.1} \text{ A}$$

Since the system is balanced, the other two currents can be obtained by noting that they lag $\hat{I}_a$ by 120° and 240°, respectively.

$$\hat{I}_b = 63.5 \; \underline{/-53.1 - 120} = 63.5 \; \underline{/-173.1} \text{ A}$$

$$\hat{I}_c = 63.5 \; \underline{/-53.1 - 240} = 63.5 \; \underline{/-53.1 + 120} = 63.5 \; \underline{/+66.9} \text{ A}$$

In addition, because this is a wye connection, these expressions also describe the line currents.

Normally, there is no need to identify the individual phase currents, and a shorter form of solution may be stated as:

$$\hat{V}_{ph} = 317.5 \; \underline{/0} \text{ V}$$

$$\hat{Z}_{ph} = 3 + j4 = 5 \; \underline{/53.1} \; \Omega$$

$$\hat{I}_{ph} = \frac{317.5 \; \underline{/0}}{5 \; \underline{/53.1}} = 63.5 \; \underline{/-53.1} \text{ A}$$

This would normally complete the solution. If the individual phase currents are required, the reference phase can be assigned at this stage. If the voltage for phase a is taken as reference, as before, this last expression for the phase current is easily interpreted as:

$$\hat{I}_a = 63.5 \; \underline{/-53.1} \text{ A}$$

$$\hat{I}_b = 63.5 \; \underline{/-53.1 - 120} = 63.5 \; \underline{/-173.1} \text{ A}$$

$$\hat{I}_c = 63.5 \; \underline{/-53.1 - 240} = 63.5 \; \underline{/-53.1 + 120} = 63.5 \; \underline{/+66.9} \text{ A}$$

Example 2-2

Obtain expressions for the phase and line currents for a balanced load when connected in delta to a balanced three-phase source of sequence abc.

Solution. In the case of a delta connection it is usually more convenient to use one of the line voltages as the reference phasor. We shall use the voltage between lines a and b since the sequence is abc. The circuit is shown in Fig. 2-7.

$$\hat{V}_{ab} = V \angle 0$$

$$\hat{V}_{bc} = V \angle -120$$

$$\hat{V}_{ca} = V \angle +120$$

The phase currents are:

$$\hat{I}_{ab} = \frac{\hat{V}_{ab}}{\hat{Z}} = \frac{V \angle 0}{Z \angle \beta} = I \angle -\beta$$

$$\hat{I}_{bc} = \frac{\hat{V}_{bc}}{\hat{Z}} = \frac{V \angle -120}{Z \angle \beta} = I \angle -\beta -120$$

$$\hat{I}_{ca} = \frac{\hat{V}_{ca}}{\hat{Z}} = \frac{V \angle +120}{Z \angle \beta} = I \angle -\beta +120$$

The line currents are:

$$\hat{I}_a = \hat{I}_{ab} - \hat{I}_{ca}$$

$$\hat{I}_b = \hat{I}_{bc} - \hat{I}_{ab}$$

$$\hat{I}_c = \hat{I}_{ca} - \hat{I}_{bc}$$

Note. The sum of the line currents, $\hat{I}_a + \hat{I}_b + \hat{I}_c$, is zero. Although all three currents are taken as positive for flow into the delta-connected load, in fact at any instant during a period at least one of the lines is providing the return path for the other(s).

If the impedance of the delta-connected load is $(30 + j40)$ Ω per phase and

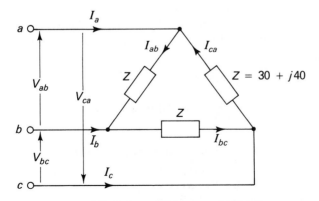

Figure 2-7   Balanced delta-connected load.

the source voltage is 550 V, the numerical values are as follows:

$$\hat{V}_{ab} = 550 \ \underline{/0} \ \text{V}$$

$$\hat{Z} = 30 + j40 \quad = 50 \ \underline{/53.1} \ \Omega$$

$$\hat{I}_{ab} = \frac{550 \ \underline{/0}}{50 \ \underline{/53.1}} \quad = 11 \ \underline{/-53.1} \ \text{A}$$

$$\hat{I}_{bc} = \frac{550 \ \underline{/-120}}{50 \ \underline{/53.1}} = 11 \ \underline{/-173.1} \ \text{A}$$

$$\hat{I}_{ca} = \frac{550 \ \underline{/-240}}{50 \ \underline{/53.1}} = 11 \ \underline{/-293.1} \ \text{A}$$

$$\hat{I}_a = \hat{I}_{ab} - \hat{I}_{ca}$$

$$= 11 \ \underline{/-53.1} - 11 \ \underline{/-293.1} = 19.05 \ \underline{/-83.1} \ \text{A}$$

As with the wye connection, it is usually sufficient to obtain this result by working with the reference phase only. That is,

$$\hat{V}_{\text{ph}} = 550 \ \underline{/0} \ \text{V}$$

$$\hat{I}_{\text{ph}} = \frac{550 \ \underline{/0}}{50 \ \underline{/53.1}} = 11 \ \underline{/-53.1} \ \text{A}$$

$$\hat{I}_{\text{line}} = \hat{I}_a = \sqrt{3} \times 11 \ \underline{/-53.1-30} = 19.05 \ \underline{/-83.1} \ \text{A}$$

The angle of 30° is subtracted from the phase angle because line currents in a balanced delta lag the phase currents by 30°.

2.4 POWER IN THREE-PHASE CIRCUITS

The average rate at which energy is dissipated or transferred in a three-phase circuit is the sum of the average power in each of the three phases. As far as each phase is concerned, the standard expression, Eq. (2-4), is valid and therefore the total power in a balanced load is simply three times the phase power:

$$P = 3 \ V_{\text{ph}} I_{\text{ph}} \cos \beta$$

$$= 3 \ \frac{V}{\sqrt{3}} I \cos \beta \ \text{if wye-connected}$$

$$= 3 \ V \frac{I}{\sqrt{3}} \cos \beta \ \text{if delta-connected}$$

$$= \sqrt{3} \ VI \cos \beta \tag{2-18}$$

where V_{ph} and I_{ph} are the phase voltage and phase current and V and I are the line voltage and line current.

Note. The phase angle in *all* of these expressions is that between *phase* voltage and *phase* current. The final expression, Eq. (2-18), is valid whatever the connection, provided both the source and the load are balanced.

Measurement of power in three-phase circuits is normally performed using standard wattmeters. However, only two are required, since this measurement does not depend on either source or load being balanced. A detailed derivation will now be given.

The basic definition of power is represented by

$$P = \frac{1}{T} \int vi \, dt \tag{2-19}$$

where v and i are the instantaneous values of voltage and current, and T is the period of the source voltage.

If three wattmeters are connected as shown in Fig. 2-8, the sum of the three readings is given by

$$P = \frac{1}{T} \int (v_{ao}i_a + v_{bo}i_b + v_{co}i_c) \, dt \tag{2-20}$$

but

$$v_{ao} = v_{am} + v_{mo}$$

$$v_{bo} = v_{bm} + v_{mo}$$

$$v_{co} = v_{cm} + v_{mo}$$

where v_{mo} is the voltage between the common points of the load and the three voltage coils of the wattmeters.

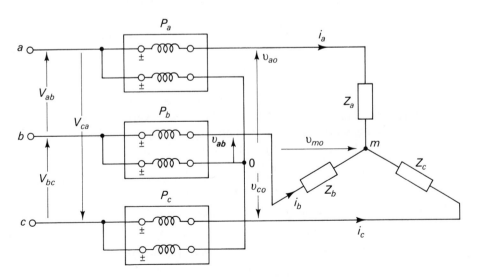

Figure 2-8 Power measurement by three wattmeters.

The sum of the readings is therefore

$$P = \frac{1}{T}\left[\int (v_{am}i_a + v_{bm}i_b + v_{cm}i_c)\,dt + \int v_{mo}(i_a + i_b + i_c)\,dt\right]$$

$$= \frac{1}{T}\int (v_{am}i_a + v_{bm}i_b + v_{cm}i_c)\,dt \quad \text{since} \quad i_a + i_b + i_c = 0 \qquad (2\text{-}21)$$

$$= \text{the total power}$$

Thus, the sum of the three wattmeter readings is equal to the total power, irrespective of the potential of the point o. Usually the point o is one of the lines, so that one of the wattmeters becomes redundant and the power is then measured using the two-wattmeter method, as shown in Fig. 2-9.

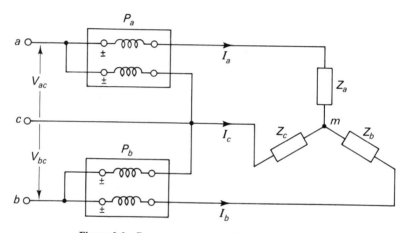

Figure 2-9 Power measurement by two wattmeters.

In terms of rms values of voltage and current, expressions for the two wattmeter readings are

$$P_a = V_{ac}\,I_a \cos (\text{angle between } \hat{V}_{ac} \text{ and } \hat{I}_a) \qquad (2\text{-}22)$$

$$P_b = V_{bc}\,I_b \cos (\text{angle between } \hat{V}_{bc} \text{ and } \hat{I}_b) \qquad (2\text{-}23)$$

When the source and load are both balanced, expressions for the two wattmeter readings are obtained by referring to the phasor diagram shown in Fig. 2-10. Using these phase angles in Eqs. (2-22) and (2-23),

$$P_a = VI \cos (30 - \beta) \qquad (2\text{-}24)$$

$$P_b = VI \cos (30 + \beta) \qquad (2\text{-}25)$$

The sum of the two readings is

$$P_s = P_a + P_b$$

$$= VI\,[\cos (30 - \beta) + \cos (30 + \beta)]$$

$$= \sqrt{3}\,VI \cos \beta$$

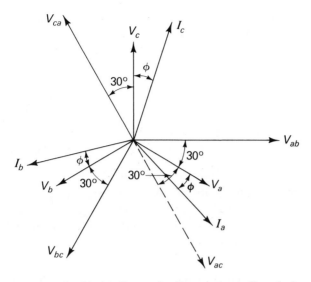

Figure 2-10 Phasor diagram for "two-wattmeter" method.

which is the same as Eq. (2-18). The difference of the two readings is

$$P_d = P_a - P_b$$

$$= VI \left[\cos (30 - \beta) - \cos (30 + \beta) \right] \qquad (2\text{-}26)$$

$$= VI \sin \beta$$

The phase angle may be obtained by noting that the ratio of the difference to the sum of the readings is

$$\frac{P_d}{P_s} = \frac{\tan \beta}{\sqrt{3}} \qquad (2\text{-}27)$$

Note. The angle β is positive for an inductive load, and thus P_a is greater than P_b for this situation, but P_a is less than P_b for a capacitive load. In addition, when the angle is 60° (*i.e.,* a power factor of 0.5), one of the wattmeters indicates zero. This is of extreme importance when taking measurements, since a wattmeter may be carrying its maximum permitted voltage and current and yet show little or no power. In addition, when the power factor is less than 0.5, one of the wattmeters has a genuine negative reading. Normally, the connections to the current coil are reversed in order to get an upscale reading, but the value *must* be interpreted as negative.

At this point it is convenient to demonstrate one of the main advantages of three-phase power transmission. The power in a single-phase circuit when expressed as a function of time is:

$$p(t) = vi$$

$$= 2VI \cos \omega t \cos (\omega t - \beta) \qquad (2\text{-}28)$$

$$= VI \cos \beta + VI \cos (2 \omega t - \beta)$$

where the first term is recognized as the average value and the second is a pulsating component that is present in all single-phase circuits. When the same approach is applied to the three-phase situation, the resulting expression is

$$p(t) = 2V_{ph}I_{ph} \cos \omega t \cos (\omega t - \beta)$$
$$+ 2V_{ph}I_{ph} \cos (\omega t - 120) \cos (\omega t - 120 - \beta) \quad (2\text{-}29)$$
$$+ 2V_{ph}I_{ph} \cos (\omega t + 120) \cos (\omega t + 120 - \beta)$$

Using the common trigonometric identities, this expression can be simplified to

$$p(t) = 3V_{ph}I_{ph} \cos \beta \quad (2\text{-}30)$$

That is, the power flow is absolutely independent of time, even with a reactive load. One result is that the torque produced by a three-phase motor is normally constant, and the motor does not subject the system being driven to pulsating torques.

Example 2-3

A balanced load of $(4 + j3)$ Ω per phase is connected in wye to a balanced 550-V three-phase source, sequence abc. Determine

(a) the three line currents,
(b) the total power dissipated,
(c) the readings of two wattmeters whose current coils are connected in lines a and b and their common voltage connection is to line c.

Solution.   The circuit diagram is shown in Fig. 2-11.

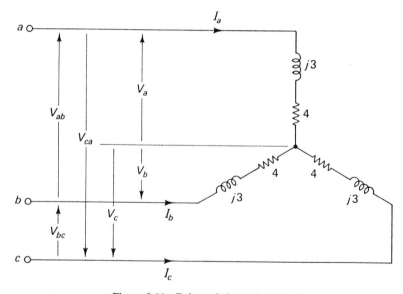

Figure 2-11   Balanced three-phase load.

(a) Since the load and source are both balanced, it is permissible to note that the magnitude of the phase voltages is $\dfrac{550}{\sqrt{3}}$. Taking $\hat{V}_a$ as reference, the phase voltages are:

$$\hat{V}_a = 317.5 \, \underline{/0} \text{ V}$$

$$\hat{V}_b = 317.5 \, \underline{/-120} \text{ V}$$

$$\hat{V}_c = 317.5 \, \underline{/+120} \text{ V}$$

The impedance of each phase is $(4 + j3)$ Ω or $5 \, \underline{/36.9}$, and therefore the currents are:

$$\hat{I}_a = \frac{317.5 \, \underline{/0}}{5 \, \underline{/36.9}} \quad = 63.5 \, \underline{/-36.9} \text{ A}$$

$$\hat{I}_b = \frac{317.5 \, \underline{/-120}}{5 \, \underline{/36.9}} = 63.5 \, \underline{/-156.9} \text{ A}$$

$$\hat{I}_c = \frac{317.5 \, \underline{/+120}}{5 \, \underline{/36.9}} = 63.5 \, \underline{/+83.1} \text{ A}$$

(b) The power in each phase is simply obtained as:

$$P = I^2 R = 63.5^2 \times 4 = 16\ 129 \text{ W}$$

and this must be the same for each phase since we are dealing with a balanced system. The total power is therefore:

$$P = 16\ 129 \times 3 = 48\ 387 \text{ W}$$

Alternatively, the total power can be obtained using:

$$P = \sqrt{3}VI \cos 36.9 = \sqrt{3} \times 550 \times 63.5 \times .8 = 48\ 393 \text{ W}$$

(c) The two wattmeter readings are:

$$P_a = V_{ac}I_a \cos (\text{angle between } \hat{V}_{ac} \text{ and } \hat{I}_a)$$
$$= 550.0 \times 63.5 \times \cos (-30 - (-36.9))$$
$$= 34\ 672 \text{ W}$$

$$P_b = V_{bc}I_b \cos (\text{angle between } \hat{V}_{bc} \text{ and } \hat{I}_b)$$
$$= 550.0 \times 63.5 \times \cos (-90 - (-156.9))$$
$$= 13\ 702 \text{ W}$$

$$P_s = P_a + P_b = 48\ 374 \text{ W}$$

Note. These angles are obtained directly from the phasor diagram in Fig. 2-12. Also, the sum of the two wattmeter readings is not quite the same as the total power calculated above. This is due to rounding off the values, especially those of phase angles. Students are encouraged to repeat these calculations using the full accuracy of their calculators, even though the large number of significant figures is not physically justified. When this is done the two calculations are in agreement.

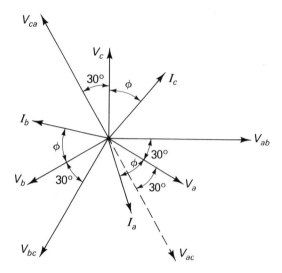

Figure 2-12   Phasor diagram, Example 2-3.

Example 2-4

Three loads are connected to a 208-V, 3-ph source having sequence *abc*. One is a three-phase motor connected in wye and running such that the line current is 25 A and the power factor is 0.8 (lagging). The second is a single-phase motor connected between lines *a* and *b* which takes a current of 8 A at a power factor of 0.707 (lagging). The third is a single-phase load connected between lines *b* and *c*, taking a current of 10 A at a power factor of 0.8 (leading).

Using $\hat{V}_{ab}$ as reference, determine

(a) all voltages,

(b) the line currents,

(c) the total input power.

Solution.   The first part of the solution is to draw the circuit diagram shown in Fig. 2-13.

(a) The voltages are:

$$\hat{V}_{ab} = 208 \ \underline{/0} \ V \qquad \hat{V}_{bc} = 208 \ \underline{/-120} \ V \qquad \hat{V}_{ca} = 208 \ \underline{/+120} \ V$$

$$\hat{V}_{a} = 120 \ \underline{/-30} \ V \qquad \hat{V}_{b} = 120 \ \underline{/-150} \ V \qquad \hat{V}_{c} = 120 \ \underline{/+90} \ V$$

(b) The currents for the first load are:

$$\hat{I}_{a1} = 25 \ \underline{/-30-36.9} \ = 25 \ \underline{/-66.9} \ A$$

$$\hat{I}_{b1} = 25 \ \underline{/-150-36.9} = 25 \ \underline{/-186.9} \ A$$

$$\hat{I}_{c1} = 25 \ \underline{/+90-36.9} \ = 25 \ \underline{/+53.1} \ A$$

where 36.9° is the angle between phase voltage and phase current for a power factor of 0.8.

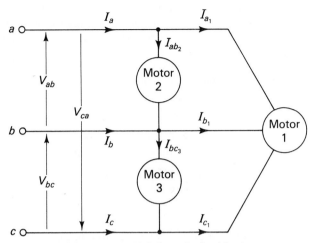

Figure 2-13 Unbalanced mixed load.

The current for the second load is

$$\hat{I}_{ab2} = 8 \; \underline{/0-45} = 8 \; \underline{/-45} \text{ A}$$

Where $0°$ is the phase angle of $\hat{V}_{ab}$ and $45°$ is the angle between the voltage and current for a power factor of 0.707. It is negative because the power factor is lagging. Since the angle for $\hat{V}_{ab}$ is used, the current $\hat{I}_{ab2}$ flows from line a to line b.

The current for the third load is

$$\hat{I}_{bc3} = 10 \; \underline{/-120+36.9} = 10 \; \underline{/-83.1} \text{ A}$$

The angle $36.9°$ comes from the power factor of 0.8, but is positive because the power factor is leading. Also, having used the phase angle for $\hat{V}_{bc}$, the current which results is $\hat{I}_{bc3}$, flowing from line b to line c.

Using Kirchhoff's current law at each of the junctions, the total line currents are:

$$\hat{I}_a = \hat{I}_{a1} + \hat{I}_{ab2}$$
$$= 25 \; \underline{/-66.9} + 8 \; \underline{/-45}$$
$$= 32.56 \; \underline{/-61.64} \text{ A}$$

$$\hat{I}_b = \hat{I}_{b1} + \hat{I}_{ba2} + \hat{I}_{bc3}$$
$$= \hat{I}_{b1} - \hat{I}_{ab2} + \hat{I}_{bc3}$$
$$= 25 \; \underline{/-186.9} - 8 \; \underline{/-45} + 10 \; \underline{/-83.1}$$
$$= 29.3 \; \underline{/-177.52} \text{ A}$$

$$\hat{I}_c = \hat{I}_{c1} + \hat{I}_{cb3}$$
$$= \hat{I}_{c1} - \hat{I}_{bc3}$$
$$= 25 \; \underline{/53.1} - 10 \; \underline{/-83.1}$$
$$= 32.95 \; \underline{/65.22} \text{ A}$$

In all of the phasor sums and differences shown above, only the polar form of the complex numbers has been shown. Although it is necessary to convert to rectangular coordinates for complex addition and subtraction, the final answer must be expressed in polar form for ease of comprehension.

(c) The power may be obtained in several ways, but it is important to recognize that the total load is not balanced and therefore any expression assuming this condition may be used only with the first of the three loads.

$$P = \sqrt{3} \times 208 \times 25 \times 0.8 \quad \text{(load 1)}$$
$$+\ 208 \times 8 \times 0.707 \qquad \text{(load 2)}$$
$$+\ 208 \times 10 \times 0.8 \qquad \text{(load 3)}$$
$$=\ 7205.3\ +\ 1176.5\ +\ 1664$$
$$=\ 10\ 045.8\ \text{W}$$

Example 2-5

An unbalanced delta-connected load is connected to a balanced 100-V three-phase source, sequence abc. The impedances are:

$$\hat{Z}_{ab} = 5 \quad\ \ + j8.66\ \Omega$$
$$\hat{Z}_{bc} = 8.66 + j5\ \Omega$$
$$\hat{Z}_{ca} = 5 \quad\ \ - j8.66\ \Omega$$

Determine the phase and line currents, and the total power dissipated in the load.

Solution. As in the previous example, it is most convenient to take the voltage between lines a and b as reference, as shown in Fig. 2-14.

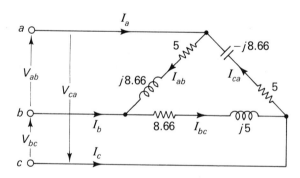

Figure 2-14   Unbalanced delta-connected load.

The phasor expressions for the three line voltages are therefore:

$$\hat{V}_{ab} = 100\ \underline{/0}\ \text{V}$$
$$\hat{V}_{bc} = 100\ \underline{/-120}\ \text{V}$$
$$\hat{V}_{ca} = 100\ \underline{/+120}\ \text{V}$$

The phase currents are then given by

$$\hat{I}_{ab} = \frac{100 \angle 0}{5 + j8.66} = \frac{100 \angle 0}{10 \angle 60}$$

$$= 10 \angle -60 = 5 - j8.66 \text{ A}$$

$$\hat{I}_{bc} = \frac{100 \angle -120}{8.66 + j5} = \frac{100 \angle -120}{10 \angle 30}$$

$$= 10 \angle -150 = -8.66 - j5 \text{ A}$$

$$\hat{I}_{ca} = \frac{100 \angle +120}{5 - j8.66} = \frac{100 \angle +120}{10 \angle -60}$$

$$= 10 \angle +180 = -10 + j0 \text{ A}$$

The line currents are:

$$\hat{I}_a = \hat{I}_{ab} - \hat{I}_{ca}$$
$$= 5 - j8.66 + 10 = 15 - j8.66$$
$$= 17.32 \angle -30 \text{ A}$$

$$\hat{I}_b = \hat{I}_{bc} - \hat{I}_{ab}$$
$$= -8.66 - j5 - 5 + j8.66 = -13.66 + j3.66$$
$$= 14.14 \angle 165 \text{ A}$$

$$\hat{I}_c = \hat{I}_{ca} - \hat{I}_{bc}$$
$$= -10 + 8.66 + j5 = -1.34 + j5$$
$$= 5.18 \angle 105 \text{ A}$$

The total power is the sum of the three phase powers:

$$P_{ab} = 10^2 \times 5 \quad = \quad 500 \text{ W}$$
$$P_{bc} = 10^2 \times 8.66 = \quad 866 \text{ W}$$
$$P_{ca} = 10^2 \times 5 \quad = \quad 500 \text{ W}$$
$$\text{Total power} = 1866 \text{ W}$$

2.5 POWER FACTOR CORRECTION

A problem that frequently arises is to ensure that the power factor of an industrial plant is sufficiently high to minimize the losses in the supply system. Any load where the power factor is less than unity implies a current that is greater than the minimum required to transmit a given power. The result is unnecessarily high losses in the transmission and distribution systems. Since the prices charged for electrical energy are determined by measurement at the consumer's terminals, the cost of these losses falls on the utility. In order to reduce these losses, typical tariffs encourage industrial customers to increase their power factor to values of the order of 0.95; the most common solution is to install capacitor banks in parallel with such loads, which are normally inductive. Usually it is not economical to raise the power factor any higher.

For the majority of industrial applications the power supply system and loads are reasonably well balanced, and calculations are performed using the equivalent single-phase circuit. A most convenient consequence of the balanced nature of the load is that generally either an equivalent wye or delta connection may be used, since the impedance values of the equivalent delta connection are three times those of the equivalent wye connection. This is certainly the case when dealing with only a simple parallel connection. However, if the series impedance of the supply is to be included in the calculations, it is usually simpler to use the equivalent wye connection, since line and phase currents are the same.

Example 2-6

When a balanced inductive load is connected to a 550-V, 60-Hz, 3-ph supply, the line current is 20 A and the total power is 10 kW. A delta-connected capacitor bank of 40 μF/ph is connected in parallel. For the parallel combination, determine

(a) the magnitude of the resulting line current,
(b) the power,
(c) the power factor.

Solution.   Since the three capacitors are connected in delta, it may be better to consider the inductive load to be delta-connected as shown in Fig. 2-15. In practice it is likely to be a combination of several loads, some of which are delta-connected and some wye-connected. If a line-to-line voltage is taken as reference, the phase

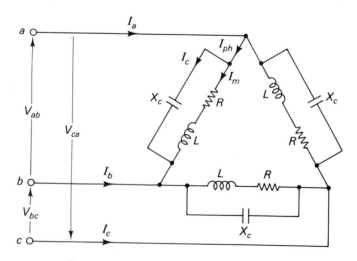

Figure 2-15   Circuit diagram for Example 2-6.

current may then be determined in magnitude and phase. Recalling that power in a balanced three-phase circuit is

$$P = \sqrt{3} VI \cos \beta$$

the power factor is readily obtained as

$$\cos \beta = \frac{10 \times 10^3}{\sqrt{3} \times 550 \times 20} = 0.5249 \text{ (lagging)}$$

and the corresponding reactive factor is

$$\sin \beta = 0.8512$$

The magnitude of the phase current is $20/\sqrt{3}$, so that the phasor representing it is

$$\hat{I}_m = 11.55 \, (0.5249 - j0.8512)$$

$$= 6.063 - j9.831$$

The reactance of each phase of the capacitor bank is

$$X_c = \frac{-j}{2\pi \times 60 \times 40 \times 10^{-6}} = -j66.31 \, \Omega$$

The phase current in the capacitor bank is thus

$$\hat{I}_c = \frac{j550}{66.31} = j8.294$$

(a) The total current is the sum of these two currents,

$$\hat{I}_{ph} = 6.063 - j9.831 + j8.294 = 6.063 - j1.537$$

$$= 6.255 \, \underline{/-14.23} \text{ A}$$

The magnitude of the total line current is $\sqrt{3}$ times this value, namely 10.835 A. Note that this is less than the current in the existing load, and since the losses in the supply line conductors are proportional to the square of the current, this would reduce these losses to 29.3% of the value corresponding to a line current of 20 A.

(b) Since the capacitors are assumed to be ideal, the average power dissipated in them is zero and the total power is the same as before.

$$P = 10 \text{ kW}$$

(c) The simplest way to determine the power factor is to evaluate the cosine of the angle between phase current and phase voltage. That is,

$$\text{Power factor} = \cos \, (-14.23°)$$

$$= 0.969 \text{ (lagging)}$$

Alternatively, the resulting apparent power may be evaluated as 10.321 kVA and the power factor obtained as $\frac{10}{10.321} = 0.969$.

2.6 NEGATIVE PHASE SEQUENCE

In all the discussion and examples so far, the sequence has been *abc*. However, this is not always the case. The same simplicity of the calculations involving positive phase sequence may be obtained by recalling the cyclic nature of the subscripts.

In the case of a wye connection, it will be more convenient to use the line-to-neutral voltage as reference. Referring to Fig. 2-1b, the three voltages are then:

$$\hat{V}_a = \frac{V}{\sqrt{3}} \angle 0$$

$$\hat{V}_c = \frac{V}{\sqrt{3}} \angle -120$$

$$\hat{V}_b = \frac{V}{\sqrt{3}} \angle +120$$

The corresponding line voltages are obtained as before in Sec. 2-3 and are:

$$\hat{V}_{ac} = V \angle 30$$

$$\hat{V}_{cb} = V \angle -90$$

$$\hat{V}_{ba} = V \angle +150$$

These three voltages lead the corresponding line-to-neutral voltages by 30°. However, if the line voltages normally labelled for sequence abc had been used (that is, $\hat{V}_{ab}$, $\hat{V}_{bc}$, $\hat{V}_{ca}$), such negative-sequence line voltages would lag the line-to-neutral voltages by 30°.

In the case of a delta connection, the line voltage will normally be used as reference, and the three voltages are:

$$\hat{V}_{ac} = V \angle 0$$

$$\hat{V}_{cb} = V \angle -120$$

$$\hat{V}_{ba} = V \angle +120$$

As long as the normal polarity convention is adhered to, there is no difficulty in dealing with negative phase sequence. The phase currents in a delta-connected load would flow from a to c, from c to b, and from b to a. The line current entering the vertex a from line a is $\hat{I}_{ac} - \hat{I}_{ba}$, with the other two being the corresponding differences in the appropriate phase currents.

PROBLEMS

2-1. Three identical impedances of $(8.66 + j5)$ Ω are connected in wye to a 460-V balanced three-phase source. Determine:

(a) the magnitude of the line currents,

(b) the total power dissipated for the three phases,

   (c) the readings of two wattmeters when connected for the standard two-wattmeter method of measuring power. The current coils are connected in lines b and c, and the sequence is abc.

2-2. Repeat Prob. 2-1 for the case where the three impedances are connected in delta.

2-3. Three identical impedances of $(12 + j15)$ Ω are connected in wye to a 550-V balanced three-phase source. Determine:
   (a) the magnitude of the line currents,
   (b) the total power dissipated for the three phases,
   (c) the total reactive power,
   (d) the power factor.

2-4. Repeat Prob. 2-3 for the case where the three impedances are connected in delta.

2-5. Current, voltage, and power to a balanced three-phase circuit are measured and found to be 20 A, 550 V, and 10.5 kW, respectively. Determine equivalent circuits per phase as follows:
   (a) wye-connected, series combination of resistance and reactance in each phase,
   (b) delta-connected, parallel combination of resistance and reactance in each phase.

2-6. Voltage, apparent power, and power to a balanced three-phase circuit are measured and found to be 460 V, 50 kVA, and 48.5 kW respectively. Determine equivalent circuits per phase as follows:
   (a) wye-connected, parallel combination of resistance and reactance in each phase,
   (b) delta-connected, series combination of resistance and reactance in each phase.

2-7. The current, voltage, and power factor of a balanced three-phase circuit are measured and found to be 15 A, 440 V, and 0.75 lagging respectively. Determine equivalent series-connected resistance and reactance circuits per phase if the phases are
   (a) wye-connected,
   (b) delta-connected.

2-8. The voltage, apparent power, and power factor of a balanced three-phase circuit are measured and found to be 600 V, 150 kVA, and 0.9 leading respectively. Determine equivalent parallel connected resistance and reactance circuits per phase if the phases are
   (a) wye-connected,
   (b) delta-connected.

2-9. Three impedances are connected in delta to a balanced 208-V, three-phase source of sequence abc. The impedances are

$$\hat{Z}_{ab} = 10 + j20 \ \Omega$$

$$\hat{Z}_{bc} = 20 - j10 \ \Omega$$

$$\hat{Z}_{ca} = 20 + j10 \ \Omega$$

(a) What are the three phase voltages?

(b) Calculate the three phase currents.

(c) Calculate the three line currents.

(d) Calculate the readings of a two-wattmeter measurement of power where the current coils are connected in lines a and c.

(e) Calculate the sum of the phase powers and compare with the sum of the two wattmeter readings.

2-10. Repeat Prob. 2-9 for the situation where the sequence is acb.

2-11. Three impedances are connected in delta to a balanced 460-V, three-phase source of sequence abc. The impedances are

$$\hat{Z}_{ab} = 25 + j15 \ \Omega$$

$$\hat{Z}_{bc} = 15 - j25 \ \Omega$$

$$\hat{Z}_{ca} = 20 + j20 \ \Omega$$

(a) What are the three phase voltages?

(b) Calculate the three phase currents.

(c) Calculate the three line currents.

(d) Calculate the readings of a two-wattmeter measurement of power where the current coils are connected in lines b and c.

(e) Calculate the sum of the phase powers and compare with the sum of the two wattmeter readings.

2-12. Two loads are connected to a 208-V, 3-ph balanced source, sequence abc. One is a 3-ph motor connected in delta and running such that the line current is 10 A with a power factor of 0.866 (lagging). The other is a single-phase heater which takes a current of 15 A at a power factor of 0.98 (lagging) when connected between lines a and b. Using $\hat{V}_{ab}$ as reference, determine the three line currents.

2-13. Two loads are connected to a 460-V, 3-ph balanced source, sequence abc. One is a 3-ph balanced load connected in wye and having a line current of 25 A with a power factor of 0.955 (lagging). The other is a single-phase load which has a current of 18 A at a power factor of 0.75 (leading) when connected between lines a and c. Using $\hat{V}_{ab}$ as reference, determine the three line currents.

2-14. Two loads are connected to a 460-V, 3-ph balanced source, sequence abc. One is a 3-ph motor connected in delta and running such that the power is 20 kW with a line current of 30 A. The power factor is known to be lagging. The other is a single-phase 10-kW heater that takes a unity PF current of 22 A when connected between lines b and c. Using $\hat{V}_{ab}$ as reference, determine the three line currents.

2-15. Two inductive loads are connected to a 460-V, 3-ph balanced source, sequence abc. One is a 3-ph balanced load of 50 kW connected in wye and having a line current of 125 A. The other is a single-phase load of 5 kW and 10 kVA connected between lines a and c. Using $\hat{V}_{ab}$ as reference, determine the three line currents.

2-16. A 3-ph induction motor is connected to a balanced 460-V, 60-Hz supply. For a particular mechanical load the current is 50 A and the power is 30 kW. The power factor is to be increased to 0.95 (lagging) by means of a wye-connected capacitor bank connected to the motor terminals. Determine the capacitance per phase required.

2-17. A 3-ph induction motor is connected to a balanced 550-V, 60-Hz supply. For a particular mechanical load the input is 100 kVA and 80 kW. The power factor is to be increased to 0.95 (lagging) by means of a delta-connected capacitor bank connected to the motor terminals. Determine the capacitance per phase required.

2-18. When a certain 3-ph induction motor is operated at its rated load the current, voltage, and power are 70 A, 550 V, and 50 kW respectively. A second motor, when connected to the same source, takes a current of 50 A and a power of 30 kW. Normally both motors operate simultaneously.
  (a) Determine the delta-connected capacitance per phase required to raise the power factor to 0.95 (lagging).
  (b) With this value of capacitance remaining in the circuit, determine the resulting power factor when the second motor is disconnected.

2-19. A certain inductive, balanced 3-ph load dissipates 50 kW with a current of 55 A when connected to a 550-V, 60-Hz supply.
  (a) Obtain the parameters of the equivalent wye-connected circuit in which the reactance and resistance are connected in series.
  (b) A set of three capacitors, each 500 μF, is connected in series with the load. Determine the current, voltage, and power of the original load.
  (c) Obtain the overall power factor.

2-20. A balanced 3-ph, wye-connected load consists of a resistance of 5 Ω connected in series with an inductance of 10 mH in each phase. The line-to-neutral voltage of phase a is

$$v_{an}(t) = 141 \cos 377t$$

  (a) Determine the line current, $i_a(t)$.
  (b) Obtain an expression for the instantaneous power dissipation in this phase.
  (c) Obtain an expression for the instantaneous total 3-ph power dissipation.

2-21. A balanced 3-ph, delta-connected load consists of a resistance of 5 Ω connected in series with an inductance of 10 mH in each phase. The voltage between lines a and b is

$$v_{ab}(t) = 141 \cos 377t$$

  (a) Determine the line current, $i_a(t)$.
  (b) Obtain an expression for the instantaneous power dissipation in the phase ab.
  (c) Obtain an expression for the instantaneous total 3-ph power dissipation.

3

MAGNETIC CIRCUITS

3.1 INTRODUCTION

Transformers and most electric machines operate on the principle of the magnetic field and are dependent on the properties of ferromagnetic materials. Because of their high relative permeability, it is possible to use lumped circuit models based on dc circuit analysis. Inevitably they involve approximations which limit their accuracy, but experience has shown that they are sufficiently accurate for most design work. However, it should be noted at the outset that magnetic circuits are nonlinear and therefore the analogous electric circuits are also nonlinear, corresponding to the situation where the resistance is a function of the current.

3.2 THE SERIES MAGNETIC CIRCUIT

There are three expressions that are used frequently to model magnetic fields in ferromagnetic cores. The first relates the magnetic flux to flux density.

$$\phi = \oint \mathbf{B} \cdot d\mathbf{A} \tag{3-1}$$

where $\mathbf{B}$ is the flux density in teslas (webers per square meter) and $\mathbf{A}$ is the cross-sectional area in square meters. This is simply a statement that the total flux is the sum of the components normal to the area through which the flux is considered to pass. If the field is uniform and normal to the area under consideration, a situation that prevails in most of the devices under study, Eq. (3-1) simplifies to

$$\phi = B A \tag{3-2}$$

The second relates the flux density, **B**, to the *magnetizing force,* **H**.

$$\mathbf{B} = \mu\mathbf{H} = \mu_r\mu_o\mathbf{H} \tag{3-3}$$

where **H** is in amperes per meter and μ is the permeability of the material. For convenience, μ is taken as the product of two components: μ_o, the permeability of free space ($= 4\pi \times 10^{-7}$) and μ_r, the relative permeability of the material.

The third expression relates the magnetomotive force, or mmf ($\mathscr{F}$), to magnetizing force.

$$\mathscr{F} = Ni = \oint \mathbf{H} \cdot d\mathbf{l} \tag{3-4}$$

If the circuit consists of different sections, each of which has a uniform cross section, this becomes

$$\mathscr{F} = \sum H_i l_i \tag{3-5}$$

This is analogous to Kirchhoff's voltage law for the summation of voltages (electromotive forces) around a closed path. The different parts of a magnetic circuit are in series if the flux passing through each part is the same. This is the identical criterion to that for the series electric circuit.

Example 3-1

A magnetic ring is shown in Fig. 3-1. Obtain an expression for the flux produced by the coil in terms of the dimensions of the ring, the number of turns, and the current.

If the mean radius of the core is 10 cm, the cross-sectional area of the core is 2.0 cm^2, there are 500 turns in the coil, and the relative permeability of the core is 1500, determine the flux produced by a current of 1.0 A in the coil.

Solution.   In practice, the coil must be wound as evenly as possible around the ring. If the ring is "narrow," that is, the radial width of the core is much less than the mean radius of the core, the flux distribution may reasonably be taken as uniform. The (magnetic) length of the complete core equals $2\pi r$, since this is the path of the

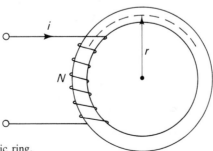

Figure 3-1   Magnetic ring.

magnetic flux. The mmf is given by

$$\mathcal{F} = Hl = H\,(2\pi r)$$
$$= \text{The current enclosed by the flux path around the ring}$$
$$= Ni$$

Hence

$$H = \frac{Ni}{2\pi r}$$

and since

$$B = \mu_r \mu_o H = \frac{\phi}{A}$$

the flux is

$$\phi = \frac{A\mu_r \mu_o Ni}{2\pi r}$$

Substituting the numerical values in this expression gives the flux as 0.3 mWb.

Since mmf is analogous to voltage, the ratio of mmf to flux is analogous to resistance in a dc circuit and is called the *reluctance* of the magnetic

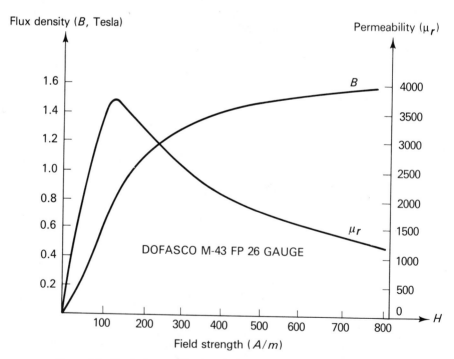

Figure 3-2 Typical magnetization characteristic of electrical sheet steel.

circuit. That is,

$$\mathcal{R} = \frac{\mathcal{F}}{\phi} \tag{3-6}$$

$$= \frac{l}{\mu_r \mu_o A} \tag{3-7}$$

Inspection of a typical magnetization characteristic such as Fig. 3-2 shows that flux density is generally not proportional to magnetizing force, and therefore flux is generally not proportional to mmf. As a result, the reluctance is not constant and therefore is usually of little value in calculations other than for an air gap such as that considered in the next example. It is, however, useful in obtaining a general idea of how a magnetic circuit behaves.

Example 3-2

The ring shown in Fig. 3-3 has a small radial cut in it, thus giving a magnetic circuit with two parts in series. If the core has the same dimensions as the one in Example 3-1 and the gap is 1.0 mm, determine the flux produced by a current of 1.0 A.

Solution.   In this case, the flux path is common to the two parts. Since the flux in the core is the same as that in the gap, the two parts are magnetically in series. The mmf expression becomes:

$$\mathcal{F} = Ni = H_g l_g + H_c l_c$$

$$= \phi_g \mathcal{R}_g + \phi_c \mathcal{R}_c \text{ (in terms of reluctance)}$$

$$= \phi (\mathcal{R}_g + \mathcal{R}_c) \text{ since } \phi_g = \phi_c$$

$$= \phi \left[\frac{l_g}{\mu_o A} + \frac{l_c}{\mu_r \mu_o A} \right]$$

From which

$$\phi = \frac{\mu_o A Ni}{l_g + \dfrac{l_c}{\mu_r}}$$

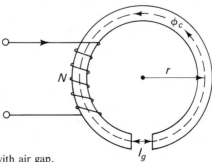

Figure 3-3   Magnetic ring with air gap.

Using the numerical values, the length of the core (excluding the air gap) is (628.3 − 1.0) mm. The error in neglecting the reduction of the core due to the gap in the above expression is evidently very small.

The width of the air gap is its magnetic length, that is

$$l_g = 0.001 \text{ m}$$

and the value of the reluctance, $\mathscr{R}_g$, is 3.98 × 10^6 A/Wb.

Substituting the same relative permeability, 1500, for the core gives a value of 1.67 × 10^6 A/Wb for its reluctance, $\mathscr{R}_c$.

The equivalent reluctance of the complete magnetic circuit is therefore 5.65 × 10^6 A/Wb, and the flux produced by a current of 1.0 A in the coil is 0.0885 mWb.

Notes

1. If the relative permeability of the core is significantly greater than the ratio l_c/l_g, the effect of the air gap dominates the characteristic of the circuit. Often this dominance is sufficient to justify neglecting the core in the same way that the conductors in an electric circuit are assumed to have zero resistance. However, in the case of a magnetic circuit, there is usually significant error. Nevertheless, it is common to assume that the effect of the core is negligible as a first approximation, and this simplifies considerably the explanation of the properties of many electro-mechanical devices.

2. If the reluctance of the core is neglected, the calculated value of the flux would be 0.126 mWb.

3. The air-gap flux has been assumed to pass directly across the gap. Normally this is correct only near the center of the gap, but at the outer edges there is a bulging effect called *fringing*. If it is not possible to plot the magnetic field in detail, it is common to apply an empirical correction to the dimensions of the gap, resulting in a cross-sectional area slightly larger than the magnetic core at each side of the gap. The most common correction is to add the length of the gap to the width and breadth when calculating its effective area. If the cross section of this core were a rectangle, 1 cm by 2 cm, its effective area using this approximation would be 2.31 cm^2. However, it must be noted that the accuracy of this correction depends on the relative sizes of the gap dimensions.

4. The unit of reluctance has been given as amperes per weber (A/Wb). Occasionally the reciprocal henry (H^{-1}) is used.

At this point it is useful to tabulate the analogous variables. In Table 3-1 it may be seen that some concepts used frequently in linear circuit analysis are of limited value in magnetic circuits, which are often saturated. The analogous terms used more frequently in magnetic circuit analysis are, of course, of considerable value when analyzing nonlinear circuits.

TABLE 3-1   Analogous Quantities

Electrical	Magnetic
Current, i (A)	Flux, ϕ (Wb)
Resistance, R (Ω)	Reluctance, $\mathscr{R}$ (A/Wb)
Voltage (Emf), E (V)	Mmf, $\mathscr{F}$ (A)
Conductivity, σ ($\Omega \cdot$m)	Permeability, μ (H/m)
Current density, J (A/m^2)	Flux density, B (T)
Voltage gradient (V/m)	Magnetizing force, H (A/m)

3.3 THE PARALLEL MAGNETIC CIRCUIT

Two sections of a magnetic core may be considered to be in parallel if the magnetic potential drop (mmf drop) is the same for both sections and the flux divides between them. The division of flux is dependent on the relative values of the reluctances, just as the division of current between two parallel branches is dependent on the relative values of the resistances. Normally it is not possible to have a magnetic circuit with only two sections that are in parallel, and therefore this situation will be illustrated using the series-parallel configuration shown in Fig. 3-4.

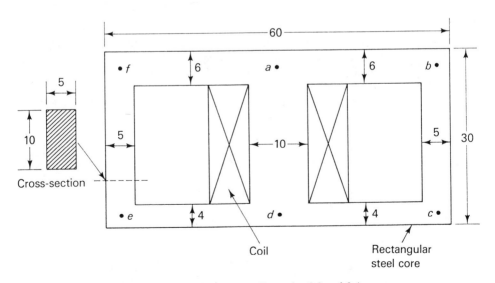

Figure 3-4   Rectangular core, Examples 3-3 and 3-4.

Example 3-3

The magnetic core shown in Fig. 3-4 has a coil of 500 turns wound on its center limb. All dimensions shown are in centimeters. The cross section of each part is rectangular. The core is made of steel laminations with a stack height of 10 cm and a stacking factor of 0.95. Determine the current required to produce a flux density of 1.2 T in the center limb, assuming a relative permeability of 3000.

Solution. Since the relative permeability is assumed to have a constant value of 3000, the reluctance concept may be employed usefully. In any event, it is still a very powerful method of determining which, if any, sections of the circuit are in series and which, if any, are in parallel.

It is usually helpful to sketch the flux path. In this way it is evident whether flux is common to two sections or whether it divides between them. This gives the analogous electric circuit shown in Fig. 3-5.

For clarity, it is convenient to label each corner of any rectangular core whether or not the flux divides. In this way the reluctance of each part may be identified clearly. Common practice is to take the length of each section as the distance between the intersections of the center lines of each part. Using this procedure, the reluctance of the path ab is

$$\mathcal{R}_{ab} = \frac{0.275}{3000 \times \mu_o \times 0.06 \times 0.1 \times 0.95}$$

$$= 12\ 798 \text{ A/Wb}$$

Similarly, the remaining reluctances are calculated and have the following values:

$$\mathcal{R}_{ab} = 12\ 798 \qquad \mathcal{R}_{bc} = 13\ 961 \qquad \mathcal{R}_{cd} = 19\ 196$$

$$\mathcal{R}_{de} = 19\ 196 \qquad \mathcal{R}_{ef} = 13\ 961 \qquad \mathcal{R}_{fa} = 12\ 798$$

$$\mathcal{R}_{ad} = \ \ 6\ 980.5$$

The path $abcd$ consists of three sections where the flux is the same in each part; that is, they form a series connection. The equivalent reluctance is therefore the sum of the three reluctances, 45 955. The path $defa$ forms a similar series connection, and because of the symmetry the equivalent reluctance of this path is also 45 955. The mmf drop across these two paths is the same, and they therefore form a parallel

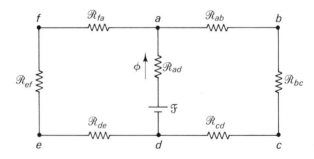

Figure 3-5   Analogous electric circuit, Example 3-3.

connection. The equivalent reluctance of this parallel connection is

$$\mathscr{R} = \frac{45\ 955 \times 45\ 955}{(45\ 955 + 45\ 955)} = 22\ 977.5 \text{ A/Wb}$$

The center limb is in series with the parallel connection formed by the two outer limbs; thus, the equivalent reluctance of the complete circuit is

$$\mathscr{R}_e = 22\ 977.5 + 6\ 980.5 = 29\ 958 \text{ A/Wb}$$

The flux density desired in the center limb is 1.2 T (tesla), so the flux is:

$$\phi = 1.2 \times 0.1 \times 0.1 \times 0.95 = 0.0114 \text{ Wb}$$

and the mmf required is

$$\mathscr{F} = 0.0114 \times 29\ 958 = 341.5 \text{ A}$$

The current required is therefore $\dfrac{341.5}{500}$ or 0.683 A.

When a ferromagnetic core is operated at a flux density such that the flux is not proportional to the mmf, it is necessary to use the magnetization characteristic directly. Although the reluctance concept is of value in determining those parts of the magnetic circuit in series and those in parallel, it is of little further use in determining numerical values. The magnetization characteristic may be used graphically, but if small increments are used, linear interpolation of tabulated values is likely to produce acceptable results.

Example 3-4

If the core shown in Fig. 3-4 has the following magnetization characteristic, determine the mmf required to produce a flux density of 1.2 T in the limb bc.

B (T)	0.4	0.8	1.0	1.2	1.4	1.5	1.6	1.7	1.75
H (A/m)	10	20	25	30	40	60	95	200	500

Solution. In this case the general relation between flux and mmf must be applied for each section of the core, with mmf drops added for series-connected sections and fluxes added for parallel-connected sections.

For limb bc, the flux is $1.2 \times 0.05 \times 0.1 \times 0.95 = 5.7$ mWb, and this is the flux in sections ab and cd. The flux densities in sections ab and cd are found by dividing 0.0057 by their respective areas. From the magnetization characteristic the values of H are found, and hence the three mmf drops can be obtained:

$$ab: \quad B = 1.0, \quad H = 25, \quad l = 0.275, \quad \mathscr{F} = 6.875$$
$$bc: \quad B = 1.2, \quad H = 30, \quad l = 0.25, \quad \mathscr{F} = 7.5$$
$$cd: \quad B = 1.5, \quad H = 60, \quad l = 0.275, \quad \mathscr{F} = 16.5$$

The mmf drop along the path $abcd$ is the sum of the Hl components:

$$\mathscr{F}_{abcd} = 6.875 + 7.5 + 16.5 = 30.875 \text{ A}$$

Since the core is symmetrical, the same values hold for the other side, namely $defa$, and the total flux in the center limb, ad, is the sum of these two fluxes, 11.4 mWb. Dividing this value by the cross-sectional area of the center limb gives the flux density in this part, 1.2 T. Again using the magnetization characteristic, the magnetizing force required for this flux density is found to be 30 A/m, so that the mmf drop in the center limb is therefore 7.5 A.

The total mmf is thus given by

$$\mathscr{F} = \mathscr{F}_{abcd} + \mathscr{F}_{ad} = \mathscr{F}_{defa} + \mathscr{F}_{ad} = 38.375 \text{ A}$$

3.4 INDUCTANCE

Inductance is normally defined as the flux linkage (λ) per ampere:

$$L = \frac{\lambda}{i} = \frac{N\phi}{i} \tag{3-8}$$

This definition applies for both the self-linkages and mutual linkages produced by a current. Consider the core shown in Fig. 3-6 with n distinct coils electrically isolated from each other. The coils are linked magnetically by the flux in the core, and some of this flux generally does not link all the coils.

A number of inductances can be defined as

$$L_{jk} = \frac{\text{Flux linking coil } j \text{ due to current in coil } k}{\text{Current in coil } k}$$

or

$$L_{jk} = \frac{\lambda_{jk}}{i_k} = \frac{N_j\phi_{jk}}{i_k} \tag{3-9}$$

where ϕ_{jk} is the flux produced by the current in coil k linking coil j. The mutual flux ϕ_{jk} may be considered as

$$\phi_{jk} = K_k\phi_k \tag{3-10}$$

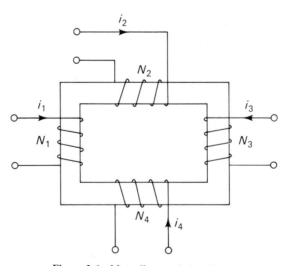

Figure 3-6   Mutually coupled coils.

where ϕ_k is the total flux produced by the current in coil k when all other currents are zero, and K_k is the *coupling factor,* which accounts for the fact that all the flux produced by the current in coil k may not link with coil j.

If $j = k$ the inductances L_{11}, L_{22}, ... L_{nn} are called *self-inductances.* All others (*e.g.,* L_{12}) are called *mutual inductances.* The mutual inductances are symmetrical such that

$$L_{jk} = L_{kj} \tag{3-11}$$

Consider the magnetic core shown in Fig. 3-7, on which are placed two coils of N_1 and N_2 turns.

$$\phi_1 = \text{Total flux linking coil 1 due to } i_1$$

$$= \phi_{21} + \phi_{\sigma 1} \tag{3-12}$$

$$\phi_2 = \text{Total flux linking coil 2 due to } i_2$$

$$= \phi_{12} + \phi_{\sigma 2} \tag{3-13}$$

where $\phi_m = \phi_{12} = \phi_{21}$ is the mutual flux linking both coils, $\phi_{\sigma 1}$ is the leakage flux linking coil 1 but not coil 2, and $\phi_{\sigma 2}$ is the leakage flux linking coil 2 but not coil 1.

The self-inductances are:

$$L_{11} = \frac{N_1 \phi_1}{i_1} \tag{3-14}$$

$$L_{22} = \frac{N_2 \phi_2}{i_2} \tag{3-15}$$

The mutual inductances are:

$$L_{12} = \frac{N_1 \phi_{12}}{i_2} = \frac{N_1 (K_2 \phi_2)}{i_2} = \frac{K_2 N_1}{N_2} \frac{N_2 \phi_2}{i_2} \tag{3-16}$$

$$L_{21} = \frac{N_2 \phi_{21}}{i_1} = \frac{N_2 (K_1 \phi_1)}{i_1} = \frac{K_1 N_2}{N_1} \frac{N_1 \phi_1}{i_1} \tag{3-17}$$

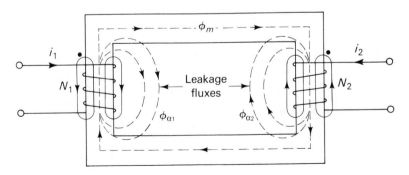

Figure 3-7   Core with leakage flux.

since $K_2 = \dfrac{\phi_{12}}{\phi_2}$ and $K_1 = \dfrac{\phi_{21}}{\phi_1}$.

From Eqs. (3-16) and (3-17)

$$L_{12} L_{21} = K_1 K_2 L_{11} L_{22}$$

Since $L_{12} = L_{21} = M$, the mutual inductance,

$$M^2 = K_1 K_2 L_{11} L_{22}$$

or

$$M = K \sqrt{L_{11} L_{22}} \qquad (3\text{-}18)$$

where $K = \sqrt{K_1 K_2}$ is the *coupling coefficient*. If there are more than two coils, there is a separate coupling coefficient for each pair of coils.

From Eq. (3-6) the self-inductance may be obtained in terms of reluctance:

$$L = \frac{N\phi}{i} = \frac{N^2 \phi}{Ni} = \frac{N^2}{\mathcal{R}} \qquad (3\text{-}19)$$

This gives the inductance directly in henrys. It should now be evident why the reciprocal henry may be taken as the unit of reluctance.

Example 3-5

The magnetic circuit shown in Fig. 3-8 has a core for which the permeability is to be taken as 1200. The coils have $N_1 = 1000$ turns and $N_2 = 500$ turns. The stacking factor is 1.0 and leakage flux is to be neglected. All dimensions shown in Fig. 3-8 are in millimeters.

(a) Draw the analogous circuit for the core, showing the values of the reluctance.
(b) Determine the self-inductance of coil 1, L_{11}.
(c) Determine the self-inductance of coil 2, L_{22}.
(d) Determine the mutual inductances, L_{12} and L_{21}.
(e) Determine the coupling coefficient.

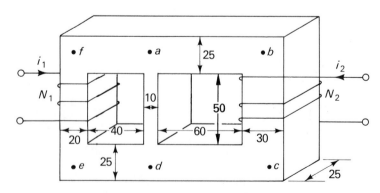

Figure 3-8 Magnetic circuit, Example 3-5.

Solution.   (a) The reluctance of each part of the magnetic circuit is obtained by substituting the appropriate values of length and area into Eq. (3-7). The resulting values are

$$l_{ab} = 8.0 \text{ cm} \qquad A_{ab} = 6.25 \text{ cm}^2 \qquad \mathcal{R}_{ab} = 0.849 \times 10^5 \text{ A/Wb}$$

$$l_{bc} = 7.5 \text{ cm} \qquad A_{bc} = 7.50 \text{ cm}^2 \qquad \mathcal{R}_{bc} = 0.663 \times 10^5 \text{ A/Wb}$$

$$l_{de} = 5.5 \text{ cm} \qquad A_{de} = 6.25 \text{ cm}^2 \qquad \mathcal{R}_{de} = 0.583 \times 10^5 \text{ A/Wb}$$

$$l_{ef} = 7.5 \text{ cm} \qquad A_{ef} = 5.00 \text{ cm}^2 \qquad \mathcal{R}_{ef} = 0.995 \times 10^5 \text{ A/Wb}$$

$$l_{ad} = 7.5 \text{ cm} \qquad A_{ad} = 2.50 \text{ cm}^2 \qquad \mathcal{R}_{ad} = 1.989 \times 10^5 \text{ A/Wb}$$

The other two values are obtained by noting that the dimensions are the same as two of the parts above. Thus

$$\mathcal{R}_{cd} = \mathcal{R}_{ab} = 0.849 \times 10^5 \text{ A/Wb}$$

$$\mathcal{R}_{fa} = \mathcal{R}_{de} = 0.583 \times 10^5 \text{ A/Wb}$$

The analogous circuit is therefore that shown in Fig. 3-9; when interpreted appropriately, it is an invaluable aid in determining the values of inductance.

(b) For a self-inductance, it is the flux linkages with the exciting coil which are required, and this is represented by the current flow through the source that is analogous to the mmf when all other sources are zero. Thus the reluctance appropriate to the determination of L_{11} is the driving point reluctance as seen from the terminals of source $\mathcal{F}_1$. That is,

$$\mathcal{R}_{11} = \left[0.995 + 0.583 \times 2 + \frac{1.989 \times 2.361}{1.989 + 2.361} \right] \times 10^5$$

$$= 3.241 \times 10^5 \text{ A/Wb}$$

Hence

$$L_{11} = \frac{N_1^2}{\mathcal{R}_{11}} = 3.085 \text{ H}$$

(c) Similarly, the self-inductance of the second coil is obtained from the driving point reluctance as seen from source $\mathcal{F}_2$, namely

$$\mathcal{R}_{22} = \left[0.663 + 0.849 \times 2 + \frac{1.989 \times 2.161}{1.898 + 2.161} \right] \times 10^5$$

$$= 3.397 \times 10^5 \text{ A/Wb}$$

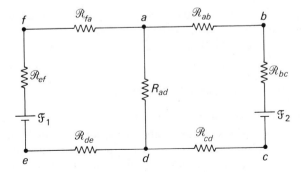

Figure 3-9   Analogous circuit, Example 3-5.

Hence

$$L_{22} = \frac{N_2^2}{\mathscr{R}_{22}} = 0.736 \text{ H}$$

(d) For the mutual inductances, we require the flux linkages with one coil produced by excitation of the other. In the analogous circuit this is the current flowing through the branch containing a suppressed source while the other is excited. First, taking excitation by coil 1, the total flux is given as above by

$$\phi_1 = \frac{\mathscr{F}_1}{\mathscr{R}_{11}}$$

and the flux linking coil 2 can be obtained using the analog of the common expression for the division of current between two parallel branches. That is,

$$\phi_{21} = \frac{1.989}{1.989 + 2.361} \phi_1 = 0.457 \phi_1$$

and

$$L_{21} = \frac{N_2 \phi_{21}}{i_1} = 0.457 \frac{N_2}{i_1} \frac{N_1 i_1}{\mathscr{R}_{11}}$$

$$= 0.457 \frac{N_1 N_2}{\mathscr{R}_{11}} = 0.705 \text{ H}$$

Similarly, for excitation by the second coil, the flux linking the first coil is

$$\phi_{12} = \frac{1.989}{1.989 + 2.161} \phi_2 = 0.479 \phi_2$$

and

$$L_{12} = \frac{N_1 \phi_{12}}{i_2} = 0.479 \frac{N_1}{i_2} \frac{N_2 i_2}{\mathscr{R}_{22}}$$

$$= 0.479 \frac{N_2 N_1}{\mathscr{R}_{22}} = 0.705 \text{ H}$$

(e) The coupling coefficient may be obtained from Eq. (3-18), since all three inductances are known. Thus

$$K = \frac{0.705}{\sqrt{(3.085 \times 0.736)}} = 0.468$$

Alternatively, the values 0.457 and 0.479 may be recognized as the coupling factors K_1 and K_2 respectively, so that the coupling coefficient can be obtained directly as

$$K = \sqrt{K_1 K_2} = \sqrt{0.457 \times 0.479} = 0.468$$

3.5 ANALYSIS OF MAGNETIC CIRCUITS

The solution of nonlinear magnetic circuit problems is straightforward, provided the implications of series and parallel connections are recognized and respected. In general, a series-connected circuit in which the flux or flux density is given may be solved quite simply by noting that flux is common to all parts, calculating the flux density in each part, obtaining the magnetizing force, H, required for each part from the magnetization characteristic, and finally summing up the Hl components to obtain the total mmf required. The analogous electric circuit problem is a series circuit excited by a current source.

Similarly, a parallel connection for which the mmf is known is solved by dividing the mmf by each length to obtain the magnetizing force in each section, obtaining the corresponding flux density from the magnetization characteristic, multiplying these values of flux density by the net cross-sectional area to obtain the respective fluxes, and adding these fluxes to obtain the total flux. The analogous electric circuit problem is a parallel circuit excited by a voltage source.

The most common difficulty is found when, for example, in a series circuit it is the mmf that is known, or in a parallel circuit it is the total flux that is known. These may require iterative solutions in which trial and error play the major part. However, in all cases it is possible to derive a flux-mmf characteristic for the complete circuit. This is done by obtaining the flux-mmf relation (usually in tabular form) for each section and then combining them. For two parts in series, sample values of flux are chosen, and for each of these the mmf components are added to give the total mmf required. For two parts in parallel, sample values of mmf are chosen, and for each of these the flux components are added to give the total flux for the two parts. This process is repeated for all the series and parallel connections until the characteristic for the complete circuit is obtained.

For the case where there is a single ferromagnetic part in series with an air gap, it is possible to obtain a solution directly, using a graphical approach. In this case

$$Ni = H_g l_g + H_c l_c$$

or

$$H_c = \frac{Ni}{l_c} - \frac{H_g l_g}{l_c} = \frac{Ni}{l_c} - \frac{B_g}{\mu_o} \frac{l_g}{l_c}$$

$$= \frac{Ni}{l_c} - \frac{B_c A_c l_g}{\mu_o A_g l_c} \qquad\qquad (3\text{-}20)$$

which has the form $y = c - mx$ and is the equation of a straight line. When this is plotted as shown in Fig. 3-10, the intersection gives the solution.

Superposition is generally *not* applicable to nonlinear circuits, and this is true for magnetic circuits. If two coils are mounted on the same limb or on two limbs that are magnetically series-connected, the total response must be obtained by adding the two mmf components, taking due note of polarity and using this as the effective mmf. Note that the resulting flux is *not* the sum of the two component fluxes calculated from the separate mmfs of the individual coils.

Flux density (*B*, Tesla)

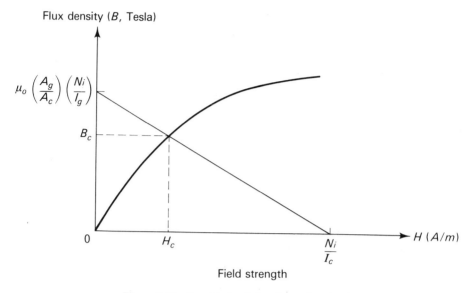

Field strength

Figure 3-10   Graphical solution for series circuit.

Example 3-6

A magnetic core similar to that of Example 3-4 and having the same magnetization characteristic is to have a flux density of 1.2 T in its limb *bc*. The dimensions shown in Fig. 3-11 are in centimeters; the only difference in this core is that the lengths of sections *de* and *fa* have been increased by 10 cm, thus making the core asymmetrical. The stack height is 10 cm and the stacking factor is 0.95, as before. Determine the mmf required.

Solution. The first part is the same as in Example 3-4, since we are dealing with a series circuit in which the flux density and hence the flux are known. The mmfs required for sections *ab*, *bc*, and *cd* are obtained in the same way and are repeated below for clarity.

$$ab: \quad B = 1.0, \quad H = 25, \quad l = 0.275, \quad \mathcal{F} = 6.875$$
$$bc: \quad B = 1.2, \quad H = 30, \quad l = 0.25, \quad \mathcal{F} = 7.5$$
$$cd: \quad B = 1.5, \quad H = 60, \quad l = 0.275, \quad \mathcal{F} = 16.5$$

The mmf drop around or along the path *abcd* is the sum of the *Hl* components, namely 30.875 A. At this point the solution diverges from that of the previous example, because the situation for path *defa* is that of a series circuit in which the mmf is known. There are two ways to attack this problem. The first is to try out a value of flux in this path, repeating the process above, and adjusting the value of the flux until the mmf for *defa* also equals 30.875 A. The other is to obtain the flux-mmf characteristic and then find the value of flux when the mmf is 30.875 A. The latter method has been chosen.

For each value of flux density in the magnetization characteristic the values of flux density in the series-connected sections are obtained. One section (in this

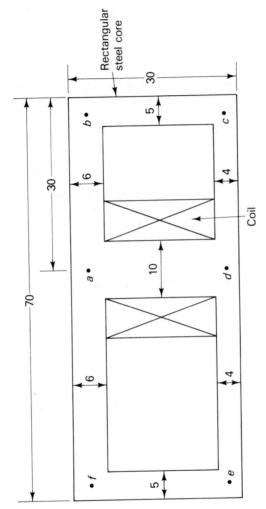

Figure 3-11 Rectangular core.

case *de)* may be chosen arbitrarily to be represented by the actual magnetization characteristic. This is reproduced below.

B_{de}	0.4	0.8	1.0	1.2	1.4	1.5	1.6	1.7	1.75
H_{de}	10	20	25	30	40	60	95	200	500

Since $A_{ef} = A_{de} \times \dfrac{5}{4}$, the values of flux density in *ef* which satisfy the continuity of flux in path *defa* are obtained by multiplying those above by the ratio 4/5. Similarly, the corresponding values of flux density in *fa* are obtained by multiplying the values for B_{de} by 4/6. These are now tabulated.

B_{ef}	0.32	0.64	0.8	0.96	1.12	1.2	1.28	1.36	1.40
B_{fa}	0.267	0.533	0.667	0.8	0.933	1.0	1.067	1.133	1.166

For each value of B_{de}, the corresponding value of flux density in each section (B_{ef} and B_{fa}) is obtained and used with the original magnetization characteristic, B_{de} versus H_{de}, to determine the value of magnetizing force for each section (H_{de}, H_{ef}, H_{fa}). These values of H are multiplied by their respective lengths to obtain the mmf drop for each section and hence for the complete path *defa*. The flux density in section *de* is multiplied by its net cross-sectional area to obtain the flux. To illustrate this process, the calculations for one point are shown in detail.

$$B_{de} \quad 1.4 \qquad H_{de} = 40 \qquad l_{de} = 0.375 \qquad \mathscr{F}_{de} = 15.0$$

$$B_{ef} = 1.12 \qquad H_{ef} = 28 \qquad l_{ef} = 0.250 \qquad \mathscr{F}_{ef} = 7.0$$

$$B_{fa} \quad 0.9333 \qquad H_{fa} = 23.333 \qquad l_{fa} = 0.375 \qquad \mathscr{F}_{fa} = 8.75$$

$$\mathscr{F}_{defa} = 15.0 + 7.0 + 8.75 = 30.75 \text{ A}$$

$$\phi_{defa} = \phi_{ef} = 1.4 \times 4 \times 10 \times 0.95 \times 10^{-4} = 5.32 \text{ mWb}$$

The complete characteristic can now be tabulated.

ϕ_{de} (mWb)	1.52	3.04	3.8	4.56	5.32	5.7	6.08	6.46	6.65
$\mathscr{F}_{defa}$	8.25	16.5	20.625	24.75	30.75	39.375	54.125	95.125	208.4

Interpolation of this characteristic gives the flux in *de* as 5.33 mWb when the mmf drop across it is 30.875 A. Adding this flux to that in *abcd*, the flux in the center limb is obtained as 11.03 mWb. When this is divided by the area of the center limb, the flux density is found to be 1.161 T, and from the original magnetization characteristic, B_{de} versus H_{de}, the magnetizing force is 29.01 A/m. Multiplying this by the length, 0.25 m, the mmf drop in the center limb is found to be 7.25 A, which when added to the 30.875 A mmf drop across both *abcd* and *defa*, gives a total of 38.125 A required from the coil.

3.6 SINUSOIDAL EXCITATION

When the exciting current is sinusoidal there is little difference from the previous sections, and the flux per ampere is determined in the same manner. However, the core is continuously subjected to changing and reversing values of magnetizing force, and the relation between flux and mmf is typically that shown in Fig. 3-12. It will be noted that the response for increasing excitation differs from that for decreasing excitation. This phenomenon is known as *hysteresis,* and the complete characteristic forms a loop or a set of loops for different peak values of excitation. If the core is saturated at the higher currents within each period, the flux waveform and hence the voltage waveform will not be sinusoidal. However, when the more common situation of sinusoidal voltage excitation prevails, the current must adjust itself to produce the necessary induced voltage. If the core is saturated, the excitation current is no longer sinusoidal and may have very pronounced peaks that depend on the peak magnetizing force demanded by the peak flux.

Since the induced voltage equals the time rate of change of flux linkages,

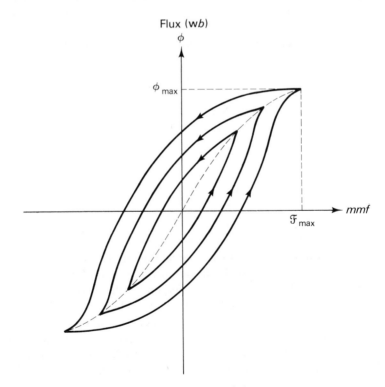

Figure 3-12   Hysteresis loop.

a sinusoidal applied voltage must produce a sinusoidally varying flux, provided the resistive voltage drop is negligible. Thus we may begin by assuming a flux having the form

$$\phi = \phi_m \sin \omega t \tag{3-21}$$

The induced voltage is therefore given by

$$e(t) = N\frac{d\phi}{dt} = \frac{d(Li)}{dt} = \omega N\phi_m \cos \omega t \tag{3-22}$$

and the rms value is

$$E = \frac{1}{\sqrt{2}} \omega N\phi_m$$

$$= 4.44 \, \phi_m fN \tag{3-23}$$

where f is the frequency in hertz. It is important to note that when a sinusoidal voltage is applied to a coil, the peak value of flux is determined by the voltage, the frequency, and the turns, since the resistive voltage drop is normally very small. The exciting current adjusts itself to produce the value of peak flux required. Thus, if the flux density with the peak flux results in a saturated core, the current must increase disproportionately during each half period to provide this flux density. This is the reason for the nonsinusoidal exciting current associated with ferromagnetic cored inductors.

When the core is unsaturated and the resistance of the coil is negligible, the peak value of the current may be obtained as

$$I_m = \frac{N\phi_m}{L} = \frac{N}{L}\frac{E}{4.44fN}$$

$$= \frac{\sqrt{2}E}{\omega L} \tag{3-24}$$

This is consistent with the reactance used in basic circuit analysis.

3.6.1 Hysteresis Loss

For each complete cycle of magnetization some energy is dissipated in the material of the core. If the hysteresis loop is a plot of flux density against magnetizing force, the energy per unit volume of the material is represented by the area enclosed by the loop. When the loop is a plot of flux against mmf, the area represents the total energy dissipated in the core. Since this is a property of the material of the core, any parameter that is used to model the hysteresis loss must be measured. The usual expression for the loss per unit volume is empirical and was first introduced by Charles Steinmetz:

$$P_h = K_h f B_{max}^z \tag{3-25}$$

where f is the frequency and B_{max} is the peak flux density. When Steinmetz first proposed this relationship, he found that the value of z was approximately 1.6, but current electrical steels have an exponent that is somewhat higher and may even exceed 2.0. The coefficient K_h also depends on the material.

3.6.2 Eddy Current Loss

The steel of which the core is made is also a conductor, although its resistivity is much higher than that of copper. The changing flux induces a voltage in the core which can be thought of as an infinite number of one-turn closed coils. Currents will therefore flow in the core, and are limited mainly by its resistance. These currents are known as *eddy currents;* they are illustrated in Fig. 3-13. In order to reduce these currents and the resulting losses, cores that are subjected to alternating excitation are usually laminated, that is, they are built of a stack of sheets of steel insulated from each other.

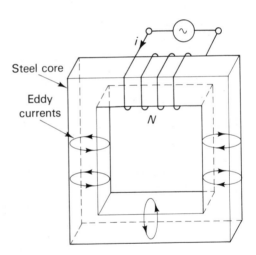

Figure 3-13 Eddy currents.

As with the hysteresis loss, the eddy current loss density must be determined experimentally and has the form

$$P_e = K_e f^2 B_{max}^2 \qquad (3\text{-}26)$$

The coefficient K_e depends on the material.

3.7 PERMANENT MAGNETS

Although hysteresis has been introduced in connection with sinusoidal excitation, the field that remains after removal of the excitation current exists in all ferromagnetic cores. For some materials this residual field is relatively large and is exploited in the form of a permanent magnet. At first sight this appears to contradict Eq. (3-4) since there is no external excitation current. However, we recall an alternate version of Eq. (3-3), namely

$$\mathbf{B} = \mu_o\,(\mathbf{H} + \mathbf{M}) \tag{3-27}$$

where **M** is the *magnetization* or average dipole moment per unit volume, and is evidently the component of the field due to the material of the core. If the core is a complete toroid such as that in Fig. 3-1, this field is, in a sense, trapped inside the core, and is of no value. The value of the normal component of this residual flux density, B_r, depends on the flux density that existed before the excitation current was removed.

The simplest form of permanent magnet is obtained by cutting a radial gap such as that in Fig. 3-3. If we consider that this core had the same flux density before the excitation current is removed, we should expect that the effect of the air gap would be to produce a residual flux density lower than B_r above. Eq. (3-5) now becomes

$$H_g l_g + H_c l_c = 0 \tag{3-28}$$

where H_c is the magnetizing force within the core. If fringing at the gap is neglected, the flux density inside the core equals that in the gap so that it is given by

$$B_c = B_g = \mu_o H_g = -\frac{l_c}{l_g}\mu_o H_c \tag{3-29}$$

This is the equation of a straight line; its intersection with the magnetization characteristic gives the operating point as shown in Fig. 3-14. The insertion of the air gap may therefore be considered as equivalent to the introduction of a negative field.

A practical point to note is that such a magnet would be excited by first placing a piece of magnetically soft iron in the air gap. This is called a *keeper*, which would be removed in normal use and replaced afterward. Another point to note is that the volume of the permanent magnet is related to that of the air gap. Assuming that the cross-sectional area of the magnet

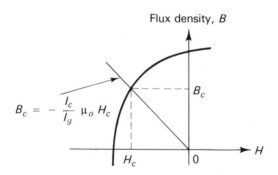

Figure 3-14 Operating point of permanent magnet.

is uniform, the volume is given by

$$A_c l_c = A_c \left(-l_g \frac{H_g}{H_c} \right) = (-) A_g l_g \frac{H_g}{H_c} \tag{3-30}$$

if fringing and leakage are neglected. This expression can be modified by recalling that the product of B and H represents the energy density within a core. If a permanent magnet is to have the maximum possible value, it is useful to multiply the numerator and denominator of Eq. (3-30) by B_g and, noting that $B_g = \mu_o H_g$, the volume of the core becomes

$$v_c = \frac{B_g^2}{\mu_o B_c H_c} v_g \tag{3-31}$$

where v_g is the volume of the air gap.

PROBLEMS

3-1. The inductor shown in Fig. P3-1 has a square cross section of 100 cm², and the coil has 1000 turns. Determine the self-inductance of the coil as a function of distance between upper and lower sections of the core, x. The reluctance of the core and fringing at the air gaps may be neglected.

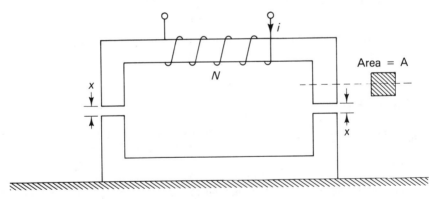

Figure P3-1

3-2. The inductor shown in Fig. P3-1 has a cross section of 100 cm². The coil has 1000 turns. The area of each air gap is 100 cm² and the length of each gap is 0.04 cm. The effective length of the complete core is 160 cm and its magnetization characteristic is as follows.

B (T)	0.12	0.22	0.42	0.70	0.96	1.2	1.31	1.40	1.45	1.55
H (A/m)	40	56	80	111	159	238	318	398	477	796

Use the graphical method to determine the flux density when the current in the coil is 1.0 A and hence determine the effective inductance.

3-3. An inductor similar to that shown in Fig. P3-1 has a coil of 800 turns, and its core has a uniform cross section of 30 cm². The length of the core is 140 cm and its magnetization characteristic is as follows.

B (T)	0.12	0.22	0.42	0.70	0.96	1.2	1.31	1.40	1.45	1.55
H (A/m)	40	56	80	111	159	238	318	398	477	796

Neglecting leakage and fringing, plot the flux-mmf characteristic when the length of each air gap is (a) 0.02 cm, (b) 0.05 cm, and (c) 0.15 cm.

3-4. The relation between flux linkages and current in an air-gap inductor may be approximated by Froelich's equation:

$$\lambda = \frac{2.25i}{0.25 + i}$$

Determine the self-inductance when the current is 2.0 A and the inductance is taken as

(a) $L = \dfrac{\lambda}{i}$

(b) $L = \dfrac{d\lambda}{di}$

3-5. The relation between flux linkages and current in an air-gap inductor may be approximated by the relation:

$$\lambda = 1.91 \, i^{0.085}$$

Determine the self-inductance when the current is 2.0 A and the inductance is taken as

(a) $L = \dfrac{\lambda}{i}$

(b) $L = \dfrac{d\lambda}{di}$

3-6. A toroidal core similar to that shown in Fig. 3-3 is to be used to provide an inductance of 0.4 H. The flux density will be low so that the relative permeability may be taken as 2000. The core has a mean radius of 10 cm and its cross section is 4 cm². If a coil of up to 1000 turns can be wound on the core, determine the width of the radial cut required, assuming that the minimum practical saw cut is 0.5 mm. Fringing at the gap is to be neglected.

3-7. A cylindrical solenoid is shown in Fig. P3-7. The plunger may move freely along its axis. The air gap between the shell and the plunger is uniform, having a radial length t of 0.25 mm. The diameter D of the plunger is 25 mm and the width w of the shell is 20 mm.

(a) Neglecting the reluctance of the core and fringing at the gaps, obtain an expression for the self-inductance of the excitation coil as a function of the length of the main gap, x.

(b) If the coil has 750 turns and carries a current of 7.5 A, what is the value of the self-inductance when $x = 2.0$ mm?

3-8. A magnetic cicuit is shown in Fig. P3-8. The air gap has an area of 0.004 m² and a length of 0.002 m. There are two coils wound on the core and interconnected as shown. Coil 1 has 600 turns, and coil 2 has 200 turns. The

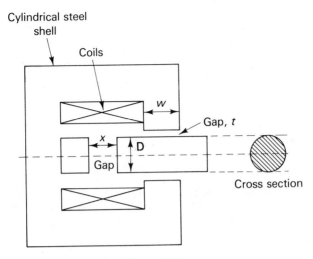

Figure P3-7

permeability of the core may be taken as infinite, and leakage and fringing at the gap may be neglected.
(a) Determine the current required to produce a flux of 0.005 Wb in the air gap.
(b) Determine the self-inductances, L_{11} and L_{22}.
(c) Determine the mutual inductances, L_{12} and L_{21}.
(d) Determine the coefficient of coupling, K.

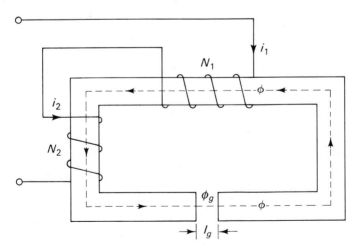

Figure P3-8

3-9. A ferromagnetic core has three limbs, in each of which there is an air gap. There is a coil mounted on each of the outer limbs, as shown in Fig. P3-9. The lengths of the gaps are $g_1 = 1.0$ mm, $g_2 = 1.5$ mm, and $g_3 = 2.0$ mm.

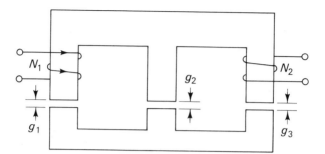

Figure P3-9

The cross section of the core at each of the three gaps is 200 mm². The reluctance of the core and fringing at the gaps are to be neglected. The coils have N_1 = 150 turns and N_2 = 450 turns. Determine
(a) the self-inductance of each winding,
(b) the mutual inductance between the two windings.

3-10. A symmetrical three-limb core, shown in Fig. P3-10, has a stack height of 100 cm and a stacking factor of 0.95. The relative permeability may be considered constant, having a value of 1500. The overall width w is 700 cm and the overall height h is 540 cm. The widths of the limbs are x_1 = 100 cm, x_2 = 100 cm, and x_3 = 100 cm. The heights of the yokes are y_1 = 120 cm and y_2 = 120 cm. The number of turns in each coil is N_1 = 100, N_2 = 100, and N_3 = 100. Determine all the self- and mutual inductances.

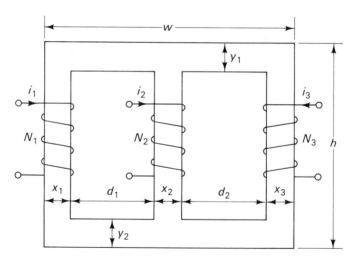

Figure P3-10

3-11. A symmetrical three-limb core, as shown in Fig. P3-10, has a stack height of 40 cm and a stacking factor of 0.9. The relative permeability may be considered constant, having a value of 2000. The overall width w is 300 cm and the overall height h is 250 cm. The widths of the limbs are x_1 = 40 cm, x_2 =

80 cm, and $x_3 = 40$ cm. The heights of the yokes are $y_1 = 50$ cm and $y_2 = 50$ cm. The number of turns in each coil is $N_1 = 200$, $N_2 = 300$, and $N_3 = 200$. Determine all the self- and mutual inductances.

3-12. A magnetic core, similar to that shown in Fig. P3-10, has a coil of 500 turns wound on one of its outer limbs. The overall width w is 50 cm and the overall height h is 30 cm. The widths of the three limbs are $x_1 = 5$ cm, $x_2 = 5$ cm, and $x_3 = 5$ cm. The heights of the yokes are $y_1 = 5$ cm and $y_2 = 5$ cm. The magnetization characteristic is given below. The stack height is 5 cm and the stacking factor is 0.95.

B (T)	0.12	0.22	0.42	0.70	0.96	1.2	1.31	· 1.40	1.45	1.55
H (A/m)	40	56	80	111	159	238	318	398	477	796

Determine the current required to produce the following flux densities:
(a) 0.4 T in the other outer limb,
(b) 0.2 T in the other outer limb,
(c) 0.4 T in the center limb,
(d) 0.2 T in the center limb.
(e) Determine the self-inductance of the coil for each of these four values of excitation.

3-13. A symmetrical magnetic core, with three limbs (Fig. P3-13), has a coil of 500 turns wound on its center limb. The overall width w is 60 cm and the overall height h is 30 cm. The widths of the three limbs are $x_1 = 5$ cm, $x_2 = 10$ cm, and $x_3 = 5$ cm. The heights of the yokes are $y_1 = 5$ cm and $y_2 = 5$ cm. The magnetization characteristic is given below. The stack height is 15 cm and the stacking factor is 0.92.

B (T)	0.127	0.233	0.445	0.742	1.018	1.273	1.389	1.485	1.538	1.644
H (A/m)	38	53	75	105	150	224	300	375	450	751

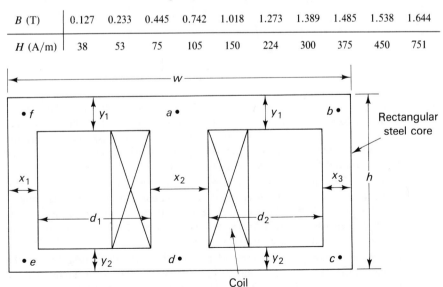

Coil

Figure P3-13

(a) Determine the current required to produce a flux density of 0.7 T in one of the outer limbs.

(b) Determine the current required to produce a flux density of 1.4 T in one of the outer limbs.

(c) Determine the current required to produce a flux density of 0.8 T in the center limb.

(d) Determine the current required to produce a flux density of 1.6 T in the center limb.

(e) Determine the self-inductance of the coil for each of these four values of excitation.

3-14. The asymmetrical core shown in Fig. 3-11 is to have a flux density of 1.0 T in the center limb. Determine the mmf required, using the same magnetization characteristic as that in Prob. 3-13 and the same dimensions as those in Example 3-6.

3-15. An air-gap inductor has a core of the form shown in Fig. P3-15. Its cross section is 5 cm² and the length of the gap is 0.25 cm. If the coil has 400 turns and is connected to a 10-V (rms), 60-Hz source, determine the peak value of the flux and of the current. If the number of turns is increased to 800, determine the peak value of the flux and of the current.

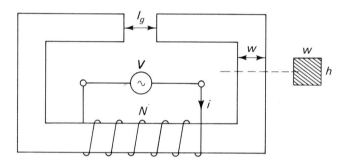

Figure P3-15

3-16. An air-gap inductor similar to that of Prob. 3-15 has 500 turns. Determine the flux density in the core when connected to a 100-V (rms), 60-Hz source. If it is to be used with a 50-Hz source having the same voltage, determine the changes required if the flux density is to remain the same. The reluctance of the core and fringing at the gap may be neglected.

3-17. An air-gap inductor such as that shown in Fig. P3-15 has a cross section of 5 cm². If the coil has 550 turns and it is connected to a 110-V, 60-Hz source, calculate the peak flux density that results. If the Steinmetz constant z is 1.6, the hysteresis loss coefficient K_h is 0.06, and the eddy current loss coefficient K_e is 0.01, calculate the core loss density.

3-18. A permanent magnet in the form of a toroid is made of material having the magnetization characteristic shown in Fig. P3-18. The mean diameter is 20 cm and the cross section is 1.0 cm². Neglecting leakage and fringing, determine the length of air gap for which the flux density is 1.0 T.

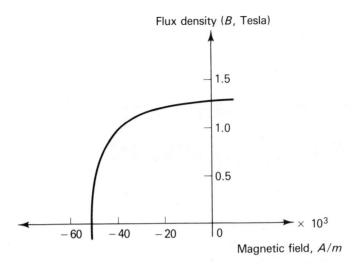

Figure P3-18

3-19. A permanent magnet in the form of a toroid is made of material having the magnetization characteristic shown in Fig. P3-18. It is to produce a flux density of 1.1 T in an air gap that is 3 cm long and has a cross section of 2.5 cm². Neglecting leakage and fringing, determine the mean diameter of the core and its volume.

4

ELECTROMECHANICAL ENERGY CONVERSION

4.1 INTRODUCTION

In Chap. 3 the relationship between flux and exciting current was extended to give a method of finding an expression for the inductance of a coil in terms of the geometry of the core. We shall now consider the energy relations associated with magnetic circuits. As far as the possibility of energy conversion is concerned, we should expect that part of the magnetic core must be able to move with respect to another part, otherwise the mechanical energy will inevitably be zero. From Chap. 3 it should be evident that inductances will generally be a function of position. When there is motion, the inductance effectively becomes a function of time even when the flux is proportional to the mmf. The expression for the voltage induced in a coil therefore becomes

$$e(t) = \frac{d}{dt}(Li)$$

$$= L\frac{di}{dt} + i\frac{dL}{dt} = L\frac{di}{dt} + i\frac{dL}{dx}\frac{dx}{dt}$$

(4-1)

We shall see that the "extra" term including dL/dx is of profound importance in the production of forces and torques.

4.2 SINGLY-EXCITED SYSTEMS

Consider the magnetic circuit shown in Fig. 4-1, where the upper part may be considered fixed and the lower part is free to move. This is not necessary, and the reverse arrangement may be considered, the final conclusions being

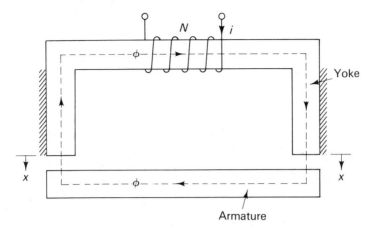

Figure 4-1   Electromagnet.

the same. The coil has N turns and the air gaps are each x meters. That is, the magnetic length of each gap is x. These two air gaps are in series; the procedure developed in Chap. 3 may be used to obtain plots of flux against mmf. We shall consider the situation at two sample positions of the movable part, or *armature*. In Fig. 4-2a the value of x is greater than that in Fig. 4-2b.

The electrical energy that goes into the coil when it is excited by a current, i, is given by

$$W_e = \int ei \, dt \tag{4-2}$$

$$= \int Ni \, d\phi = \int \mathscr{F} \, d\phi \tag{4-3}$$

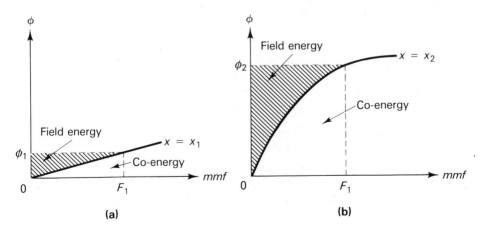

Figure 4-2   Flux-Mmf graphs.

This is always the case. When current flows through the exciting coil of a core in which no part can move, and therefore no conversion of energy is possible, Eq. (4-3) represents the energy stored in the magnetic field. It is very helpful to interpret this as the shaded areas in Fig. 4-2. If these two flux-mmf graphs are combined as in Fig. 4-3, we can now relate the electrical, magnetic, and mechanical energies.

For the linear case where flux is proportional to mmf, Eq. (4-3) can be developed to give the field energy in terms of inductance. Provided there is no motion, the energy stored in the magnetic field is

$$
\begin{aligned}
W_f = W_e &= \int i \, d(N\phi) \\
&= \int i \, d(Li) \qquad\qquad (4\text{-}4) \\
&= \int Li \, di \qquad \text{(if } L \text{ is not a function of current)} \\
&= \frac{1}{2} Li^2
\end{aligned}
$$

If flux is not proportional to mmf, the inductance concept may not be used directly, and at this stage the only approach is by means of a graphical interpretation of Eq. (4-3).

For this analysis we consider that the armature is permitted to move very slowly from $x = x_1$ to $x = x_2$ so that the current remains constant. The following expressions for the *changes* in energy are obtained.

ΔW_e = change in electrical energy   = area $abcd$

ΔW_f = change in magnetic field energy = area $ObcO$ − area $OadO$

ΔW_m = change in mechanical energy

     = $\Delta W_e - \Delta W_f$

     = area $abcd$ − area $ObcO$ + area $OadO$

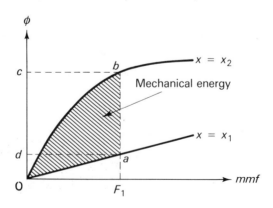

Figure 4-3 Mechanical energy as an area.

That is,

$$\Delta W_m = \text{area } OabO$$

This is simply a statement of the conservation of energy: The energy that goes into the coil from the (electrical) source must equal the sum of the increase in the energy stored in the magnetic field and the mechanical energy obtained. The area $OabO$ is shown shaded in Fig. 4-3; it represents the mechanical energy produced when the armature moves from x_1 to x_2.

The principle of virtual work is now applied to make this conclusion more useful. If the movement is very small, the motion may be considered to take place either at constant current, as above, or at constant flux. In the limit, as the movement tends to zero, the points a, b, and c in Fig. 4-4 become effectively identical. These two constraints result in two separate expressions for the force acting on the armature.

1. The armature moves with constant flux.

$$\Delta W_e = \int \mathscr{F}\, d\phi = 0 \text{ since } \phi \text{ is constant}$$
$$\Delta W_m = \text{area } OacO$$
$$= f\, \Delta x$$

where f is the force acting in the positive direction of x. Applying the conservation of energy as before, we obtain

$$\Delta W_m = \Delta W_e - \Delta W_f$$

or

$$f\, \Delta x = -\Delta W_f$$

The expression for the force becomes

$$f = -\frac{\Delta W_f}{\Delta x}$$

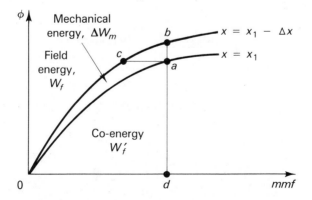

Figure 4-4   Flux-Mmf graph for incremental motion.

The negative sign results from the fact that at constant flux, all the mechanical energy must come from a decrease in the magnetic field energy. In the limit, as Δx tends to zero, if the field energy is considered as a function of flux and position, the force developed is given by

$$f = -\frac{\partial W_f(x, \phi)}{\partial x} \tag{4-5}$$

2. The armature moves with constant mmf.

In this case there is a change in the electrical energy, and in order to obtain an expression similar to Eq. (4-5), it is necessary to use the complement of the energy, or *co-energy* defined as

$$W_f' = \int \phi \, d\mathcal{F} \tag{4-6}$$

which is the area between the flux-mmf characteristic and the mmf axis. When the motion takes place at constant mmf, the change in mechanical energy is still given by the area $OabO$, which is interpreted as the change in co-energy.

$$\Delta W_m = \text{area } OdbO - \text{area } OdaO = \Delta W_f' = f \Delta x$$

Rearranging and taking the limit as before, we obtain the force developed as

$$f = \frac{\Delta W_f'}{\Delta x} \tag{4-7}$$

or

$$f = \frac{\partial W_f'(x, \mathcal{F})}{\partial x} \tag{4-8}$$

For the linear case (flux proportional to mmf), the energy and the co-energy are equal, and it is more convenient to use the inductance of the exciting coil. That is,

$$W_f' = W_f$$

and

$$f = \frac{\partial W_f'(x, \mathcal{F})}{\partial x} = \frac{\partial W_f(x, \mathcal{F})}{\partial x}$$

But

$$W_f = \frac{1}{2}Li^2 \qquad \text{(since } L \text{ is independent of current)}$$

so

$$f = \frac{d}{dx}\left[\frac{1}{2}Li^2\right]$$

$$= \frac{1}{2}i^2\frac{dL}{dx} \tag{4-9}$$

For the general case where inductances are functions of both position and current, complete solutions require an advanced mathematical background. These problems are beyond the scope of this text and they will not be considered further.

Example 4-1

The device shown in Fig. 4-1 is considered to have negligible fringing and reluctance of the core. Obtain an expression for the force developed.

Solution. The two air gaps are in series, and therefore the equivalent reluctance is twice that of one gap. That is,

$$\mathscr{R} = \frac{2x}{\mu_o A}$$

$$\phi = \frac{\mathscr{F}}{\mathscr{R}} = \frac{N i \mu_o A}{2x}$$

$$L(x) = \frac{N^2 \mu_o A}{2x}$$

$$f(x) = \frac{1}{2} i^2 \frac{dL}{dx}$$

$$= -\frac{\mu_o A N^2 i^2}{4x^2}$$

The negative sign indicates that the force acting on the armature is in the negative direction of x, namely the direction that will reduce x if the armature is free to move. This is exactly what we would expect from our observation of such magnets. For many apparently simple devices the most difficult part of the problem is that of modeling the inductance as a function of the position of the armature. It is not possible to state exactly how every problem should be tackled. All that can be done is to demonstrate by example some typical situations.

Example 4-2

A small salient-pole single-phase motor has the cross section shown in Fig. 4-5, and the reluctance of the core and fringing at the air gaps are to be neglected. The axial length of both rotor and stator is l_c. Determine an expression for the torque developed as a function of θ, the angle between the axes of rotor and stator.

Solution. It is the same flux that crosses the gaps, and they are therefore in series. The equivalent reluctance is thus twice that of one gap. When the rotor and stator are aligned, the angle θ is zero and the area of the air gaps is maximum. When the angle θ equals $2\theta_o$, the tips of the rotor are opposite those of the stator and the area is just zero. For angles greater than $2\theta_o$, the (magnetic) length of the air gaps is so large that the magnetic field may reasonably be taken as zero.

The problem is to model this information in an expression for the reluctance of the gaps. Often it is a matter of hypothesizing an expression and then examining it to see if it satisfies the conditions above. This procedure gives the reluctance as

$$\mathscr{R} = \frac{2g}{\mu_o A} = \frac{2g}{\mu_o l_c r(2\theta_o - \theta)}, 0 < \theta < 2\theta_o$$

$$\phi = \frac{Ni}{\mathscr{R}} = \frac{Ni\mu_o l_c r(2\theta_o - \theta)}{2g}$$

$$L(\theta) = \frac{N^2 \mu_o l_c r(2\theta_o - \theta)}{2g} \text{ H}$$

$$T(\theta) = \frac{1}{2} i^2 \frac{dL}{d\theta} = -\frac{N^2 i^2 \mu_o l_c r}{4g} \text{ N·m}$$

Note that the torque is developed in the direction that will align the rotor and stator—that is, the position of minimum reluctance.

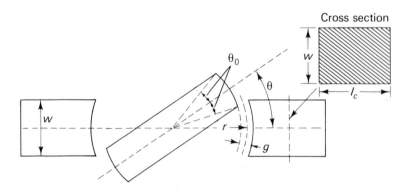

Figure 4-5   Single-phase motor cross section.

4.3 DOUBLY-EXCITED SYSTEMS

Since the expression for developed force in terms of the field energy is quite general, Eqs. (4-5) to (4-9) may be applied directly. Again, it is usually more convenient to work with the inductance, and when Eq. (4-3) is applied to a system with two excited coils the expression for the field energy becomes

$$W_f = \int (N_1 i_1 \, d\phi_1 + N_2 i_2 \, d\phi_2)$$

$$= \int [i_1 \, d(N_1\phi_1) + i_2 \, d(N_2\phi_2)]$$

$$= \int [i_1 \, d(L_{11}i_1 + L_{12}i_2) + i_2 \, d(L_{22}i_2 + L_{21}i_1)]$$

If the inductances are independent of currents, this becomes

$$W_f = \int L_{11}i_1 \, di_1 + \int L_{22}i_2 \, di_2 + \int L_{12} \, d(i_1 i_2)$$

$$= \frac{1}{2}L_{11}i_1^2 + \frac{1}{2}L_{22}i_2^2 + L_{12}i_1 i_2 \qquad (4\text{-}10)$$

and the corresponding expression for the force developed is

$$f = \frac{1}{2}i_1^2 \frac{dL_{11}}{dx} + \frac{1}{2}i_2^2 \frac{dL_{22}}{dx} + i_1 i_2 \frac{dL_{12}}{dx} \qquad (4\text{-}11)$$

Example 4-3

A small motor has a cylindrical stator on which is wound a single winding. Its rotor has salient poles and it also has a winding. These are shown schematically in Fig. 4-6. The current in the rotor is constant (dc) and that in the stator is sinusoidal (ac). Obtain an expression for developed torque as a function of currents and speed.

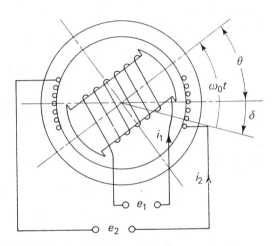

Figure 4-6   Single-phase motor with two windings.

Solution.   The periodic nature of the inductances will be explored in this problem. As before, the first part of the solution is to obtain expressions for each inductance as a function of the position of the rotor with the aid of the plots shown in Fig. 4-7. In this case we note that as far as the rotor coil is concerned, the air gap does not change as it moves around; thus, its self-inductance is independent of position. That is,

$$L_{11} = L_r \qquad (i.e., \text{ constant})$$

For the self-inductance of the stator we must note that it is a maximum when θ is 90° and 270°. It is a minimum when θ is 0° and 180° so that for each revolution of the shaft the self-inductance of the stator coil goes through *two* periods. A suitable expression to model this variation is

$$L_{22} = L_{av} - L_s \cos 2\theta$$

The mutual inductance is zero when θ is 0° and 180° and maximum when θ is 90° and 270°. However, care must be taken because the relative position between the two coils is reversed at these two positions. If the polarity of the windings is such that at $\theta = 90°$ the mutual inductance is positive, then at $\theta = 270°$ the mutual inductance must be negative. This variation is modeled by the expression

$$L_{12} = M \sin \theta$$

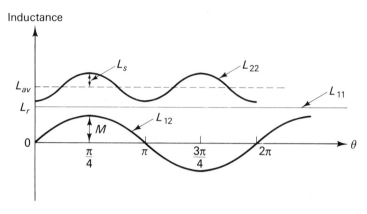

Figure 4-7 Inductance variation with position.

The currents are:

$$i_1(t) = I_1$$

$$i_2(t) = I_2 \sin \omega t$$

Let ω_o be the angular velocity of the rotor so that its position is obtained by integrating this velocity. That is,

$$\theta = \omega_o t - \delta$$

Note that δ is the position of the rotor at $t = 0$ of the current sinusoid. We are now in a position to interpret Eq. (4-11) in terms of torque and angular position. By substituting these expressions and with some algebraic manipulation, our desired expression will be obtained.

$$T(\theta) = \frac{1}{2} i_1^2 \frac{dL_{11}}{d\theta} + \frac{1}{2} i_2^2 \frac{dL_{22}}{d\theta} + i_1 i_2 \frac{dL_{12}}{d\theta}$$

$$T(\theta, t) = \frac{1}{2} I_2^2 \sin^2 \omega t \, 2L_s \sin 2\theta + I_1 I_2 \sin \omega t \, M \cos \theta$$

$$= I_2^2 L_s \sin^2 \omega t \sin 2(\omega_o t - \delta) + I_1 I_2 M \sin \omega t \cos (\omega_o t - \delta)$$

$$= \frac{I_2^2 L_s}{2} \sin 2(\omega_o t - \delta)$$

$$- \frac{I_2^2 L_s}{4} [\sin 2(\omega t + \omega_o t - \delta) + \sin 2(-\omega t + \omega_o t - \delta)]$$

$$+ \frac{I_1 I_2 M}{2} [\sin (\omega t + \omega_o t - \delta) + \sin (\omega t - \omega_o t + \delta)]$$

This is the general expression for steady-state operation of this motor. Since the average value of a sinusoidal time varying function over a complete period is zero, the average value of the developed torque is zero unless the rotor speed has a particular value, namely $\omega_o = \omega$ (or $-\omega$). This is usually called the *synchronous speed*, and the expression for torque becomes

$$T = \frac{I_2^2 L_s}{2}\left[\sin 2(\omega t - \delta) - \frac{1}{2}\sin 2(2\omega t - \delta) + \frac{1}{2}\sin 2\delta \right]$$
$$+ \frac{I_1 I_2 M}{2}[\sin (2\omega t - \delta) + \sin \delta]$$

and

$$T_{av} = \frac{I_2^2 L_s}{4}\sin 2\delta + \frac{I_1 I_2 M}{2}\sin \delta$$

Note that the motor sets the value of the angle δ as an automatic reaction to provide the torque demanded by the mechanical system so that in steady-state operation the developed torque is exactly equal to the load torque. The torque developed has a definite maximum value. Any attempt to extract more torque when motoring will result in the motor stalling. For the theoretically ideal case where there is no friction, the load torque would be zero, the angle δ would become zero, and the motor would run under this "perpetual motion" condition. If the device is generating and the load would require a torque greater than the maximum, the prime mover driving the generator will have no opposing torque and will accelerate, sometimes to a dangerously high speed.

There are two special cases of Example 4-3 that are worth commenting on at this stage.

1. If the rotor is cylindrical, L_s is zero, so that the average torque becomes

$$T_{av} = \frac{I_1 I_2 M}{2}\sin \delta$$

and the maximum or *pull-out torque* is

$$T_p = \frac{I_1 I_2 M}{2} \text{ at } \delta = 90°$$

2. If there is no coil on the rotor, the current i_1 is zero, so that the developed torque becomes

$$T_{av} = \frac{I_2^2 L_s}{4}\sin 2\delta$$

and the pull-out torque is

$$T_p = \frac{I_2^2 L_s}{4} \text{ at } \delta = 45°$$

This torque exists only because of the saliency of the rotor. If the rotor were cylindrical as discussed in (1) above, there would be no developed torque. It is therefore a direct result of the variation of reluctance with position. For this reason it is often referred to as the *reluctance torque*, which is the mechanism by which torque is developed in the *reluctance motor*.

PROBLEMS

4-1. For the electromagnet shown in Fig. P4-1 find the force developed, $f_m(x)$, and the energy required to lift the upper half from the position $x = 1.0$ mm to $x = 4.0$ mm. Assume that the relative permeability of the core is infinite; neglect magnetic fringing. The coil has 325 turns and carries a current of 4.0 A.

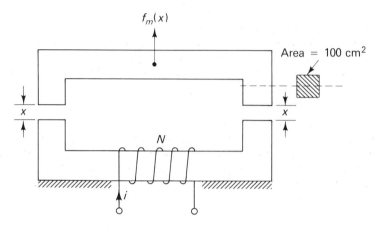

Figure P4-1

4-2. The magnet shown in Fig. P4-1 has a uniform cross section of 100 cm². Its coil has 325 turns and carries a current of 4.0 A. The mean length of the core is 260 cm; leakage and fringing may be neglected. The magnetization characteristic of the core material is as follows.

B (T)	0.12	0.22	0.42	0.70	0.96	1.2	1.31	1.40	1.45	1.55
H (A/m)	40	56	80	111	159	238	318	398	477	796

(a) Plot the flux–mmf characteristic for gaps of $x = 1.0$ mm and $x = 4.0$ mm.

(b) From these graphs, determine the change in magnetic field energy, coenergy, and mechanical energy if the armature is allowed to move from $x = 1.0$ mm to $x = 4.0$ mm with constant current.

4-3. Repeat Prob. 4-2 if the movement takes place with the flux constant at the value when $x = 1.0$ mm.

4-4. An air-gap inductor shown in Fig. P4-4 has a stack height of 2.5 cm, stacking factor $= 0.95$, $w = 2.5$ cm, and gap length $l_g = 0.1$ cm. The coil has 400 turns and carries a current of 5 A. Determine the force of attraction between both sides of the gap, neglecting the reluctance of the core, leakage, and fringing.

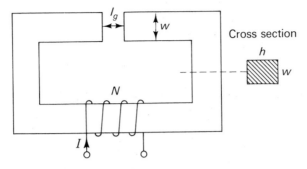

Figure P4-4

4-5. The plunger in the cylindrical solenoid shown in Fig. P4-5 is supported so that it may move freely along its axis. The diameter of the plunger, D, is 25 mm. The air gap, t, between the shell and the plunger, is uniform and has a radial length of 0.25 mm and $w = 20$ mm. The exciting coil has 750 turns. Neglecting the reluctance of the steel core and fringing at the air gaps, obtain an expression for the self-inductance of the exciting coil as a function of the length of the main air gap, x. If the exciting coil carries a current of 7.5 A, determine the force acting on the plunger when $x = 2.0$ mm.

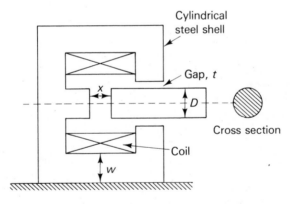

Figure P4-5

4-6. The cylindrical solenoid of Prob. 4-5 carries a constant current of 5 A; its armature is allowed to move slowly from a position at $x = 5.0$ mm to another at $x = 2.0$ mm. Calculate
(a) the change in energy stored in the magnetic field,
(b) the change in the magnetic co-energy,
(c) the change in the energy supplied by the source.

4-7. A doubly-excited electromechanical system is constrained to move horizontally. The pertinent dimensions are shown in Fig. P4-7. Winding resistance, magnetic

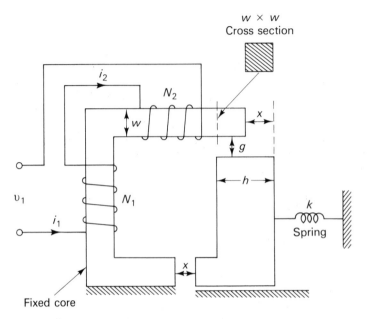

Figure P4-7

leakage, and fringing may be assumed negligible. The permeability of the core is very large compared to that of free space. Determine

(a) the reluctance of the magnetic circuit,

(b) the self-inductance of coil 1, L_{11},

(c) the self-inductance of coil 2, L_{22},

(d) the mutual inductances L_{12} and L_{21},

(e) the magnetic energy stored in the air gap,

(f) the force extended on the movable part as a function of its position.

(g) If $N_1 = 100$ turns, $N_2 = 500$ turns, $i_1 = 2$ A, $i_2 = 2$ A, $x = 0.5$ cm, $w = 2.0$ cm, $g = 0.5$ cm, and $h = 5$ cm, calculate

(1) the magnetic energy stored,

(2) the force on the movable part.

4-8. A clapper magnet is illustrated in Fig. P4-8. The inductance of the coil has the approximate relation

$$L(x) = \frac{K}{x}$$

where x is in meters. Obtain an expression for the average force exerted on the clapper as a function of x if the coil is connected to a 60-Hz sinusoidal voltage source, rms value V. The velocity of the clapper is to be considered negligible at all values of x. The resistance of the coil is also to be neglected.

Calculate the value of the force when $K = 0.004$, $V = 100$, and $x = 2.0$ cm.

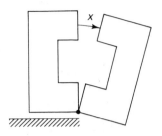

Figure P4-8

4-9. Two coils, one mounted on a stator and the other on a rotor, have self- and mutual inductances as follows.

$$L_{11} = 0.5 \text{ H}$$

$$L_{22} = 0.3 \text{ H}$$

$$L_{12} = 0.35 \cos \theta \text{ H}$$

where θ is the angle between the axes of the coils. The coils are connected in series and carry a current

$$i(t) = 20 \sin \omega t \text{ A}$$

(a) Derive an expression for the instantaneous developed torque as a function of the angular position θ.

(b) From this, obtain an expression for the (time) average torque as a function of θ.

(c) Calculate the numerical value of the average torque when $\theta = 90°$.

4-10. A small electromechanical device consists of a rotor and a stator, on each of which a coil has been wound. The inductances are:

$$L_{rr} = 1.0$$

$$L_{ss} = 2.0 \cos 4\theta$$

$$L_{rs} = 1.4 \cos 2\theta$$

where all values are in henrys and θ is the angle between the axes of the two coils.

(a) Find an expression for the torque developed as a function of the two currents, and the angle θ.

(b) If the currents, in amperes, are

$$i_r(t) = 5 \cos \omega t$$

$$i_s(t) = 2$$

find the (time) average of the developed torque when the rotor is stationary and the angle θ is 60°.

4-11. A small salient-pole single-phase motor has the cross section shown in Fig. P4-11; the reluctance of the core and fringing at the air gaps are to be neglected. The axial length of both rotor and stator is l_c and the mean diameter at the

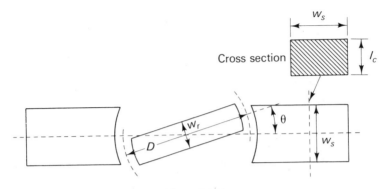

Figure P4-11

gap is D. Determine an expression for the torque developed as a function of
θ, the angle between the axes of rotor and stator.

    Determine the average value of the torque when the rotor is positioned
at $\theta = 30°$ and the current is 0.5 A if $w_r = 2.0$ cm, $w_s = 2.5$ cm, $D = 3.0$
cm, and $l_c = 3$ cm. The radial length of each air gap is 0.1 cm and there is
a total of 2000 turns on the stator.

4-12. A small motor has a cylindrical stator on which is wound a single winding.
Its rotor has salient poles but it does not have any winding. The arrangement
is shown schematically in Fig. P4-12. The self-inductance of the stator winding
has a maximum value of 1.5 H, a minimum value of 1.0 H, and is assumed
to vary sinusoidally with position. The current in the stator is supplied from
a 60-Hz sinusoidal source and has an amplitude of 3.0 A. Determine
(a) the pull-out torque when running at synchronous speed,
(b) the torque angle when the developed torque is 0.25 N·m.

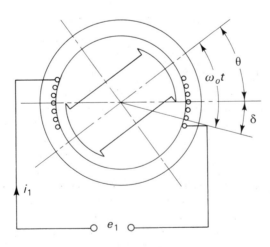

Figure P4-12

5

TRANSFORMERS

5.1 INTRODUCTION

The simplest form of transformer consists of two coupled coils. For a single-phase circuit, this changes the voltage level, with a resulting change in current level. There is no conversion of energy to or from mechanical form, but the transformer is included in this text because any electromechanical device having two or more coils is likely to have some transformer action in addition to the conversion of energy that is desired. As a result, an understanding of the transformer is necessary for the study of most motors and generators.

Figure 5-1 shows two transformer coils; the polarity of all variables is that normally used for the analysis of two-port networks. The equilibrium equations, using the Laplace transform for the normal condition of no initial currents, and neglecting the resistance of the coils, are

$$L_{11}sI_1(s) + L_{12}sI_2(s) = V_1(s) \tag{5-1}$$

$$L_{21}sI_1(s) + L_{22}sI_2(s) = V_2(s) \tag{5-2}$$

where L_{11} is the self-inductance of coil 1, L_{22} is the self-inductance of coil 2, and $L_{12} = L_{21}$ is the mutual inductance.

For power applications, the coils are mounted on a ferromagnetic core so that the coupling is high. These equations, although entirely correct and applicable for the sinusoidal steady state, do not provide the most convenient model. The most common model is based on the concept of a perfect or *ideal* transformer with its imperfections modeled by standard resistances and inductances.

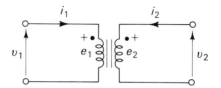

Figure 5-1   Two coupled coils.

5.2 THE IDEAL TRANSFORMER

If a transformer were truly ideal, the coupling between its two coils would be perfect (coupling coefficient equal to unity) and it would not require any current to magnetize its core (infinite permeability of the steel). That is, the flux linking the primary would be the same as that linking the secondary, as indicated in Fig. 5-2, so that the mutual flux would equal both of them.

$$\phi_1 = \phi_2 = \phi_m \tag{5-3}$$

and the corresponding rates of change of flux would also be equal. The induced voltages are given by

$$e_1 = N_1 \frac{d\phi_1}{dt} \tag{5-4}$$

$$e_2 = N_2 \frac{d\phi_2}{dt} \tag{5-5}$$

Since the rates of change of the flux are equal, the ratio of the induced voltages is given by

$$\frac{e_1}{e_2} = \frac{N_1}{N_2} = \frac{V_1}{V_2} = a \tag{5-6}$$

Where a is the *turns ratio*. For clarity, the symbol e is being used for induced voltage and v for terminal voltage. Lowercase letters are used for instantaneous values and uppercase letters are used for rms values. The

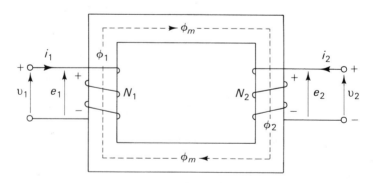

Figure 5-2   Ideal Transformer Coils.

ideal transformer thus has the property that the ratio of the terminal voltages equals the ratio of the turns.

Since the core of an ideal transformer is infinitely permeable, the net mmf must always be zero. That is,

$$N_1 i_1 + N_2 i_2 = 0 \qquad (5\text{-}7)$$

and rearranging this gives the ratio of the currents as

$$\frac{i_1}{i_2} = -\frac{N_2}{N_1} = \frac{I_1}{I_2} = -\frac{1}{a} \qquad (5\text{-}8)$$

Normally the current is taken positive in the opposite direction from that shown in Fig. 5-1, so that the negative sign disappears. This will be done when the complete circuit model is obtained in Fig. 5-6.

When an impedance is connected to the secondary winding terminals, as shown in Fig. 5-3, an expression for the input or driving point impedance is obtained by invoking the turns ratios as defined by Eqs. (5-6) and (5-8).

$$\hat{Z}_i = \hat{Z}' = \frac{\hat{V}_1}{\hat{I}_1} = \frac{\hat{E}_1}{\hat{I}_1} = \frac{N_1}{N_2} \hat{V}_2 \frac{N_1}{N_2} \frac{1}{\hat{I}_2} = a^2 \hat{Z} \qquad (5\text{-}9)$$

The usual interpretation of this expression is that the value of the load impedance, $\hat{Z}$, referred to the primary, is $a^2\hat{Z}$ or $\hat{Z}'$. Thus, the ideal transformer has the property of changing the effective value of an impedance by a factor equal to the square of the turns ratio. Another property of the ideal transformer is obtained by taking the product of voltage and current.

$$V_2 I_2 = V_1 \frac{N_2}{N_1} I_1 \frac{N_1}{N_2} = V_1 I_1 \qquad (5\text{-}10)$$

That is, the apparent power is unchanged. Moreover, since the phase angle of the input impedance is the same as that of the load impedance, the input power must therefore be the same as the output power. It should be noted that this transfer of power takes place without any direct electrical connection between the two windings. This electrical isolation is sometimes required for safety in, for example, medical equipment and in instrumentation for systems having a very high voltage. It should be evident that the power is transferred from one voltage level to the other by means of the magnetic field.

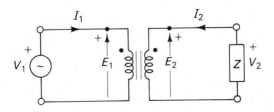

Figure 5-3 Ideal Transformer.

If the secondary winding has more turns than the primary, its voltage will be higher than that of the primary. In this case it may be called a *step-up* transformer, since its action is to raise the voltage level. For clarity, the secondary is often called the HV (high-voltage) winding, and the primary is called the LV (low-voltage) winding. If the secondary winding has fewer turns, it is called a *step-down* transformer, and the secondary is therefore the LV winding. If primary and secondary turns are equal, it is the isolating property that is being used, and it is usually then called an *isolating* transformer.

Even power transformers do not have perfect coupling, and the resistance of the windings is not negligible. The model that is derived in the next section includes an ideal transformer as one of its elements, and also has resistances and inductances to account for imperfections.

5.3 THE SINGLE-PHASE TRANSFORMER

Equations (5-1) and (5-2) will now be extended to include the resistance of each winding and rearranged to change the secondary winding variables to their values referred to the primary. Let $V_2'(s) = aV_2(s)$ be the secondary terminal voltage referred to the primary, and $I_2'(s) = I_2(s)/a$ the secondary current referred to the primary. That is, $V_2'(s)$ and $I_2'(s)$ are the primary values corresponding to $V_2(s)$ and $I_2(s)$ if the transformer were ideal. The actual values are, of course, different; and the circuit model must give the actual values.

The equilibrium equations for the actual coils are

$$(R_1 + L_{11}s)I_1(s) + L_{12}sI_2(s) = V_1(s) \tag{5-11}$$

$$L_{21}sI_1(s) + (R_2 + L_{22}s)I_2(s) = V_2(s) \tag{5-12}$$

Replacing $V_2(s)$ and $I_2(s)$ by $V_2'(s)/a$ and $I_2'(s)a$ gives

$$(R_1 + L_{11}s)I_1(s) + L_{12}s(aI_2'(s)) = V_1(s) \tag{5-13}$$

$$L_{21}sI_1(s) + (R_2 + L_{22}s)(aI_2'(s)) = \frac{V_2'(s)}{a} \tag{5-14}$$

and rearranging these equations so that they are in the standard form gives

$$(R_1 + L_{11}s)I_1(s) + aL_{12}sI_2'(s) = V_1(s) \tag{5-15}$$

$$aL_{21}sI_1(s) + (a^2R_2 + a^2L_{22}s)I_2'(s) = V_2'(s) \tag{5-16}$$

Equations (5-15) and (5-16) can now be interpreted as those of a two-mesh circuit. Since the mesh equations of the circuit shown in Fig. 5-4 are Eqs. (5-15) and (5-16), this pair of equations constitutes a circuit model of the transformer. The two inductances that appear at the top of the diagram as the difference between two terms have physical significance. Unlike a similar interpretation of Eqs. (5-1) and (5-2), these two inductances are always positive.

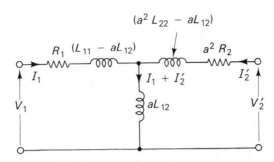

Figure 5-4   Circuit Model of Transformer.

Referring to Fig. 5-5, the mutual inductances are:

$$L_{12} = L_{21} = \frac{N_1\phi_m}{i_2} = \frac{N_2\phi_m}{i_1} \qquad (5\text{-}17)$$

The first of these difference terms can be expressed as

$$L_{11} - aL_{12} = \frac{N_1\phi_1}{i_1} - \frac{N_1}{N_2}\frac{N_2\phi_m}{i_1}$$

$$= \frac{N_1}{i_1}(\phi_1 - \phi_m) \qquad (5\text{-}18)$$

$$= \frac{N_1}{i_1}\phi_{\sigma1}$$

$$= L_{\sigma1} = L_1$$

The difference between the flux linking the primary and that linking the secondary is due to leakage. This leakage flux is illustrated in Fig. 5-5a, which shows two coils mounted on a single core. Normally, the coils would be wound on the same limb in order to minimize the leakage, but this is the simplest form for explanation. Since the permeability of the core is not infinite, its reluctance is not zero, and some of the magnetic flux passes through the open space (window) of the core. In large transformers the magnitude of the leakage is controlled by choosing the spacing between primary and secondary windings when they are being designed.

The inductance associated with the leakage flux is therefore termed *leakage inductance,* and Eq. (5-18) represents the leakage inductance of the primary. Similar manipulation of the other difference gives

$$a^2L_{22} - aL_{12} = a^2\frac{N_2}{i_2}(\phi_2 - \phi_m) \qquad (5\text{-}19)$$

$$= a^2L_{\sigma2} = a^2L_2$$

Since L_2 is the leakage inductance of the secondary, this difference represents its value, referred to the primary. These inductances can be considered as separate inductances in the primary and secondary circuits, with the

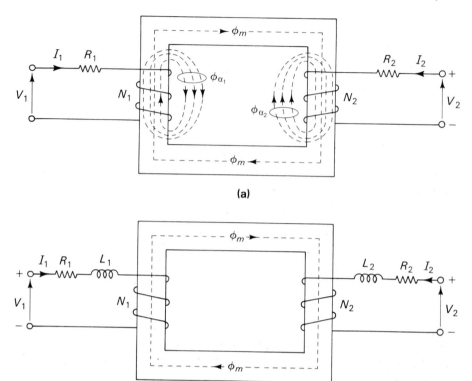

Figure 5-5 Transformer Leakage Flux.

coupling being provided by an ideal transformer. This is illustrated in Fig. 5-5b.

The third inductance shown in Fig. 5-4 is

$$aL_{12} = \frac{N_1}{N_2} \frac{N_2 \phi_m}{i_1}$$

$$= L_m$$

(5-20)

which is usually called the magnetizing inductance.

These leakage inductances are perhaps the most significant parameters in the operation of a power transformer, although the others are not really negligible. Another aspect that must be included in the equivalent circuit is the core loss, which is modeled by a resistance, R_m. In Chap. 3 it was noted that there are losses in a core which is subjected to sinusoidal excitation, and that they are dependent on the peak value of the flux density. In the circuit model of Fig. 5-3 it is the magnetizing inductance that models this flux; therefore the core loss resistance is placed in parallel with it. Note that the magnetizing inductance reflects the fact that the core of the transformer

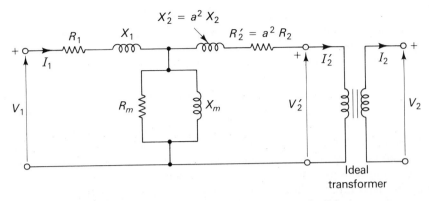

Figure 5-6 Complete Circuit Model Referred to the Primary.

is not infinitely permeable, but requires some current in order to produce the flux.

The complete circuit model referred to the primary is shown in Fig. 5-6, although it is not common to show the ideal transformer at the right-hand side. It is equally valid to refer all values to the secondary, and the alternative version of the model is shown in Fig. 5-7. This time the ideal transformer is connected to the primary, and again it is not common to show it. In accordance with normal practice, it will not be shown any further.

The model that has just been developed must be used in this form if a transformer is being used for instrumentation. However, if it is a question of a normal power transformer, it is possible to simplify the circuit so that calculations are significantly shortened without introducing significant error. It is based on the fact that in a typical power transformer the series equivalent impedance of the magnetizing branches is very much greater than those of

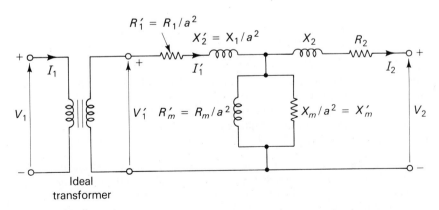

Figure 5-7 Complete Circuit Model Referred to the Secondary.

the windings. As a result, the error in the calculated value of the voltage across the primary resistance and leakage reactance is small when the approximate primary current (i.e., the secondary current referred to the primary) is used. It must be emphasized that the circuit shown in Fig. 5-8 is not an exact equivalent as this term is used in network analysis. It is simply an approximation which has been found to give results that are sufficiently close for calculations involving currents, voltages, and losses.

The sum of the two series-connected resistances is called the *equivalent resistance,* and the sum of the two inductances is the *equivalent inductance.* Since power transformers are normally operated at only one frequency, it is common to show the *equivalent reactance* on circuit models so that it may be used directly in all sinusoidal steady-state calculations.

When using the equivalent resistance and the equivalent reactance we must be careful to note that they may be referred to either the primary or the secondary. Whatever version is chosen, it is important to keep in mind that the equivalent resistance and equivalent reactance include the resistance and leakage reactance of *both* windings referred only to the chosen winding. We must therefore be very clear as to which winding the equivalent impedance is referred.

This approximate circuit model may be used to determine the *voltage regulation,* which is defined as the change in load voltage when the load is removed, usually expressed as a ratio of the load voltage.

$$\text{Regulation} = \frac{V_{oc} - V_t}{V_t} \tag{5-21}$$

where V_{oc} is the open circuit or no-load secondary voltage and V_t is the secondary terminal voltage for a particular load.

Example 5-1

A single-phase transformer has a nominal voltage ratio of 25 000/5 000 V and it has the following parameter values.

$$
\begin{aligned}
R_1 &= 100, & X_1 &= 220 \\
R_2 &= 4, & X_2 &= 10 \\
R_m &= 200\,000, & X_m &= 100\,100
\end{aligned}
$$

(Both R_m and X_m are referred to the HV winding.)

where all values are in ohms at the rated frequency of 60 Hz. The rating of the transformer is 100 kVA.

If the transformer delivers 100 kVA at 5000 V when the power factor is 0.8 (lagging), determine the primary voltage required, the voltage regulation, the losses, and the primary current.

Solution. Since we are asked to determine primary voltage and current, there is a marginal advantage in referring all parameters to the primary. Normally, the nominal voltage ratio is in fact the turns ratio, so that the secondary resistance and leakage reactance are referred to the primary by multiplying by the square of this

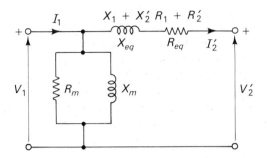

Figure 5-8   Approximate Circuit Model.

ratio. These values are shown in Fig. 5-9. The values of magnetizing resistance and reactance are already referred to the primary.

Since the output current, voltage, and power factor are given, the choice of reference phasor is between secondary terminal voltage and secondary current. When the terminal voltage is taken as reference phasor, there is surprisingly little variation in the phase angle of the primary voltage, irrespective of the size of the transformer and the power factor of the load. Having chosen this voltage as reference phasor, the load current must be obtained in phasor form. The magnitude of this current is given by

$$I_2 = \frac{100\ 000 \text{ VA}}{5000 \text{ V}} = 20 \text{ A}$$

$$I'_2 = \frac{20 \times 5000}{25\ 000} = 4 \text{ A}$$

The phasor expressions are obtained by noting that the power factor is the cosine of the phase angle between the secondary voltage and current.

$$\hat{V}'_2 = 25\ 000 + j0 \text{ V}$$

$$\hat{I}'_2 = 4(0.8 - j0.6) = (3.2 - j2.4) \text{ A}$$

The equivalent resistance, referred to the primary, is $100 + 100 = 200 \ \Omega$; the equivalent reactance, referred to the primary, is $220 + 250 = 470 \ \Omega$. With the

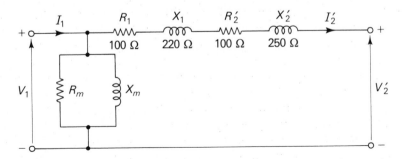

Figure 5-9   Circuit Model, Example 5-1.

simplified model, the magnetizing branches do not affect the voltage relations since they are connected directly across the source. As a result, the voltage law may be applied directly to obtain the primary voltage.

$$\hat{V}_1 = 25\ 000 + j0 + 4\ (0.8 - j0.6)\ (200 + j470)$$
$$= 26\ 768 + j1024$$
$$= 26\ 788\ \underline{/2.19}$$

As a point of interest, if the exact model of Fig. 5-6 is used, the value $26\ 857\ \underline{/2.19}$ is obtained for the primary voltage.

The voltage regulation can be obtained directly, since the open circuit voltage at the secondary is simply 26 788 divided by the turns ratio. That is,

$$\text{Regulation} = \frac{26\ 788 - 25\ 000}{25\ 000} = 0.0715 \text{ or } 7.15\%$$

Since the voltage across R_m is now known to be 26 788 V, the core loss is obtained as

$$P_c = \frac{26\ 788^2}{200\ 000} = 3588 \text{ W}$$

The copper loss is obtained directly from the equivalent resistance. Since we have the value referred to the primary, we must use the current also referred to the primary. Thus

$$P_1 = 4^2 \times 200 = 3200 \text{ W}$$

and the total loss is 6788 W.

The current through the magnetizing branches must be found and added to the referred value of secondary current in order to determine the primary current. These must be obtained relative to the reference phasor, and there are several ways to approach this. Perhaps the simplest is to get the admittance of the magnetizing branches as

$$\hat{Y}_m = \frac{1}{R_m} - j\frac{1}{X_m}$$
$$= 0.5 \times 10^{-5} - j10^{-5}$$
$$\hat{I}_m = \hat{V}_1\hat{Y}_m = (26\ 768 + j1024)(0.5 - j1) \times 10^{-5}$$
$$= (0.144 - j0.263) \text{ A}$$

Adding this to the secondary (referred) current of $(3.2 - j2.4)$ gives the primary current as

$$\hat{I}_1 = 3.344 - j2.663 = 4.27\ \underline{/-38.5} \text{ A}$$

If this transformer were operated at the same current but with a leading power factor of 0.8, the primary voltage required would be $24\ 592\ \underline{/4.63}$, which means a voltage regulation of -1.63%. Note that the voltage regulation depends on both the magnitude and power factor of the load.

5.4 MEASUREMENT OF EQUIVALENT CIRCUIT PARAMETERS

Because the simplified circuit model consists of two main parts, it is possible to obtain reasonably accurate measured values with only two standard tests. These are an *open circuit test* normally performed at rated voltage, and a *short circuit test* normally performed at rated current. Each test can be performed by exciting either winding, but with large units involving high levels of both current and voltage it may be preferable to excite the low-voltage winding on open circuit and to excite the high-voltage winding on short circuit.

Referring to Fig. 5-7, it may be seen that on open circuit there is no current flow through the equivalent impedance. This is not strictly correct, but we should keep in mind that the impedance of the shunt branches is typically of the order of 500 times that of the equivalent impedance. If V_{oc}, I_{oc}, and P_{oc} are the measured values of input voltage, current, and power on open circuit, the parameter values are obtained simply from

$$Y_{oc} = |\hat{Y}_{oc}| = \frac{I_{oc}}{V_{oc}} \tag{5-22}$$

$$\cos \beta_{oc} = \frac{P_{oc}}{I_{oc}V_{oc}} \tag{5-23}$$

The open circuit (complex) admittance for this inductive circuit is obtained as

$$\hat{Y}_{oc} = Y_{oc}(\cos \beta_{oc} - j \sin \beta_{oc})$$
$$= \frac{1}{R_m} - j\frac{1}{X_m} \tag{5-24}$$

The impedance on short circuit is usually taken as being the equivalent impedance. Again, this is an approximation that is normally valid because the series equivalent impedance of the parallel-connected magnetizing branches is usually very much greater than the equivalent impedance of the transformer. If the measured values of input current, voltage, and power on short circuit are I_{sc}, V_{sc}, and P_{sc}, the equivalent resistance and reactance are obtained from

$$Z_{eq} = |\hat{Z}_{eq}| = \frac{V_{sc}}{I_{sc}} \tag{5-25}$$

$$\cos \beta_{sc} = \frac{P_{sc}}{V_{sc}I_{sc}} \tag{5-26}$$

and the (complex) equivalent impedance of this inductive circuit is

$$\hat{Z}_{eq} = Z_{eq}(\cos \beta_{sc} + j \sin \beta_{sc}) \qquad (5\text{-}27)$$

$$= R_{eq} + jX_{eq} \qquad (5\text{-}28)$$

If the short circuit and open circuit tests have been performed by exciting different windings, it is necessary to refer one set of the measured resistance and reactance to the other winding. When performing the short circuit test at rated current, the voltage required is typically between 5% and 10% of the rated value of the winding being excited.

These standard tests do not provide enough information to separate the resistance and reactance of each winding. Normally this is not really necessary, because the relatively high impedance of the magnetizing branches makes the approximate circuit model quite acceptable. Often, in the absence of additional information, the values for the resistance and leakage reactance of one winding when referred to the other are taken as equal to the resistance and leakage reactance of the other winding.

5.5 EFFICIENCY

The efficiency of a transformer is the ratio of the output power to the input power. This definition also applies to all the machines that are studied in this text. In the case of any alternating current device, it is important to calculate the ratio of the actual powers, since the ratio of the apparent powers has no real significance. Although the efficiency can always be calculated by determining the ratio directly, there are occasions when the alternate expressions given in Eq. (5-29) are more convenient and round-off errors in calculation are less likely to cause problems.

$$
\begin{aligned}
\text{Efficiency} &= \frac{P_{out}}{P_{in}} \\
&= \frac{P_{out}}{P_{out} + P_{loss}} \qquad (5\text{-}29) \\
&= \frac{P_{in} - P_{loss}}{P_{in}} \\
&= 1 - \frac{P_{loss}}{P_{in}}
\end{aligned}
$$

In normal operation a good transformer has only a small voltage regulation and the frequency is virtually constant. It is therefore common to consider that the core loss is constant, since there is then very little variation in the peak value of the flux. It is then possible to consider the efficiency to be a function of the load current and obtain the condition for maximum efficiency. That is,

$$\text{Efficiency} = \frac{VI \cos \beta}{VI \cos \beta + I^2 R_{eq} + P_{oc}} \qquad (5\text{-}30)$$

As usual, the current, voltage, and equivalent resistance in Eq. (5-30) must be referred to the same side of the transformer. Differentiating this with respect to the current, I, and equating to zero gives the condition for maximum efficiency as

$$I^2 R_{eq} = P_{oc} \qquad (5\text{-}31)$$

That is, the efficiency is maximum when the variable loss equals the fixed loss. Depending on the load cycle that is intended for a transformer, the load at maximum efficiency may be set at less than rated load.

Example 5-2

A single-phase 200/100-V, 60-Hz, 5-kVA transformer is subjected to standard open circuit and short circuit tests with the following results.

> Open circuit (LV excited): 100 V, 4 A, 125 W
> Short circuit (HV excited): 16 V, rated current, 250 W

(a) Determine the values of the parameters of the approximate equivalent circuit referred to the HV winding.

(b) Determine the efficiency when delivering 5 kVA at 100 V at a power factor of 0.9 (lagging).

(c) Determine the load current that will give maximum efficiency for rated secondary voltage and a power factor of 0.9 (leading).

(d) Determine the efficiency for the load found in part (c).

Solution.

(a) Since the short circuit test was performed by exciting the HV winding, the values obtained from Eqs. (5-25) to (5-28) will be referred to this winding without further manipulation. Before proceeding, it is necessary to determine the rated current, since this was the value used in the test. This is simply the rated volt-amperes divided by the rated voltage.

$$I_{rated} = \frac{5000}{200} = 25 \text{ A}$$

The magnitude of the equivalent impedance is

$$Z_{eq} = \frac{16}{25} = 0.64 \ \Omega$$

The power factor for the short circuit test is $250/400 = 0.625$, so that the (complex) impedance is

$$\hat{Z}_{eq} = 0.64 \,(.625 + j.781)$$

$$= 0.4 + j0.5$$

That is,

$$R_{eq} = 0.4 \ \Omega, \ X_{eq} = 0.5 \ \Omega$$

Since the open circuit test was performed by exciting the secondary, the values for R_m and X_m will be referred to the LV winding and must be multiplied by the square of the turns ratio to obtain the desired values. The magnitude of the open circuit admittance is

$$Y_{oc} = \frac{4}{100} = 0.04 \text{ S (siemens)}$$

The power factor on open circuit is $125/400 = 0.3125$, so that the (complex) admittance is found from Eq. (5-22) to (5-24) as

$$\hat{Y}_{oc} = 0.4 \, (0.3125 - j0.95)$$
$$= (0.0125 - j0.038) \text{ S}$$

from which

$$R_m = \frac{1}{0.0125} = 80 \ \Omega$$

and

$$X_m = \frac{1}{0.038} = 26.32 \ \Omega$$

Since the turns ratio is 200/100, these values are multiplied by 4 to get the magnetizing resistance and reactance referred to the HV winding. That is,

$$R_m = 320 \ \Omega$$

and

$$X_m = 105.28 \ \Omega$$

(b) The output power for the load given is

$$P_o = 100 \times 50 \times 0.9 = 4500 \text{ W using values referred to the LV side}$$

or

$$= 200 \times 25 \times 0.9 = 4500 \text{ W using values referred to the HV side}$$

The copper or resistive loss is

$$P_r = 25^2 \times 0.4 = 250 \text{ W}$$

and the core loss is normally taken as constant and equal to the value found in the open circuit test if it was performed at rated voltage. That is,

$$P_c = 125 \text{ W}$$

The input power is

$$P_i = 4500 + 250 + 125 = 4875 \text{ W}$$

and the efficiency is $4500/4875 = 0.923$ or 92.3%.

(c) Again assuming that the core loss has the constant value of 125 W, the efficiency will be maximum when the copper loss also equals this value. That is,

$$I^2 R_{eq} = 125$$

Using the values referred to the HV winding,

$$I_1^2 = \frac{125}{0.4} = 312.5$$

so that

$$I_1 = 17.67 \text{ A}$$

or

$$I_2 = 35.34 \text{ A}$$

(d) In this case the output power is

$$P_o = 200 \times 17.67 \times 0.9 = 3180.6 \text{ W}$$

and the maximum efficiency for this power factor is 3180.6/3430.6, which equals 0.927 or 92.7%.

5.5.1 Energy Efficiency

Transformers that are operated on variable loads generally will have their maximum efficiency at less than rated load. In choosing such a transformer, it is useful to calculate the ratio of the output to input energy for a typical load cycle. This is required when comparing the contribution of the losses to the operating costs of different transformers.

The load cycle is split up into periods where the load is approximately constant, and the losses and energy for each period are calculated. It is still common to use the kilowatt-hour as the unit of energy for this purpose. If it is a daily load cycle, the total output energy and total input energy are calculated for a typical 24-hour period and the ratio is the *energy efficiency*.

Example 5-3

A single-phase, 100-kVA, 11 000/550-V transformer has a no-load loss of 2 kW at rated voltage and a copper loss of 4 kW at rated load current. It operates on a daily load cycle as follows

Time (h)	HV Current (A)	Power Factor
8	2.0	0.9 lag
10	4.5	0.8 lag
6	9.0	0.9 lead

Determine the energy efficiency for this load cycle.

Solution. The main point to recall is that the copper loss is proportional to the square of the load current. In this case the rated current of the HV winding is 100 000/11 000 = 9.091 A, so that the copper loss when the HV current is 2.0 A, as it is during the first period, is

$$P = \left(\frac{2.0}{9.091}\right)^2 \times 4 = 0.194 \text{ kW}$$

The output power during this period is

$$P_o = 11 \times 2 \times 0.9 = 19.8 \text{ kW}$$

so that the input power during the same period is

$$P_i = 19.8 + 0.194 + 2.0 = 21.99 \text{ kW}$$

The output and input energy are obtained by multiplying these values for power by the length of the period, 8 hours, to get the energy in kilowatt-hours.

This problem is best done by tabulating the calculated values as shown in Table 5-1. It is usually convenient to number the lines so that the calculations will proceed more quickly after the first column is completed. Note that whether the power factor is lagging or leading is irrelevant for this particular problem.

TABLE 5-1   Calculation of Input and Output Energy

1. Period (h)	8	10	6
2. Current (A)	2	4.5	9
3. Power factor	0.9	0.8	0.9
4. Copper loss (kW)	0.194	0.98	3.92
5. Core loss (kW)	2.0	2.0	2.0
6. Output power (kW)	19.8	39.6	89.1
7. Input power (kW) (Sum of lines 4, 5, 6)	21.99	42.58	95.02
8. Output energy (kWh) (Line 6 × line 1)	158.4	396.0	534.6
9. Input energy (kWh) (Line 7 × line 1)	175.9	425.8	570.1

The total output energy during the 24-hour period is the sum of the values in line 8 of all three columns, namely 1089 kWh. The total input energy is the sum of the line 9 values of all three columns, namely 1171.8 kWh, so that the energy efficiency is $1089/1171.8 = 0.93$ or 93%.

5.6 PER-UNIT QUANTITIES

If we examine the nameplate of a large transformer we will find, in addition to the nominal voltage ratio and the rated volt-amperes (in kVA or MVA), the equivalent impedance. However, rather than expressing the resistance and reactance in ohms, they are normally given as percentages. Typically the resistance lies between 1% and 4%, and the reactance between 3% and 10%. These are the result of the form of normalization which is used in power systems; it is often called the *per-unit system*.

In essence, there is a choice of new units for all variables. The choice of two of the variables could, in theory, be quite arbitrary, but in practice the units of voltage and current are chosen to be the rated values for the transformer. Having chosen units for two of the variables in this fashion, all others must be derived using the normal circuit relations. Thus the unit of impedance becomes the ratio of unit (rated) voltage to unit (rated) current.

The unit of power, apparent power, and reactive power becomes the product of unit (rated) voltage and unit (rated) current.

When an impedance (in ohms) is divided by this unit of impedance, the result is the value of the impedance expressed *per-unit*. As is common elsewhere, when such values are less than unity it may be convenient to multiply them by 100; they are then expressed as percentages. Thus, when a transformer has an equivalent resistance of 2%, the actual value is 2% of the ratio of rated voltage to rated current of *one* winding, referred to that winding.

Example 5-4

Determine the per-unit values of the equivalent resistance and equivalent reactance of the 25 000/5000-V, 100-kVA transformer of Example 5-1, and repeat the calculations in the per-unit system.

Solution. The first part is to determine the value of the unit of impedance—that is, the ratio of rated voltage to rated current. If we use values referred to the 25 kV winding, this is 25 000/4 = 6250 Ω.

The actual value of the equivalent resistance referred to this winding is 200 Ω, and dividing this by the unit impedance of 6250 Ω gives its normalized value as 0.032 pu or 3.2%. Similarly, since the equivalent reactance referred to the 25 kV winding is 470 Ω, its normalized value is 470/6250 = 0.0752 pu or 7.52%.

The load is 100 kVA at 5000 V and 0.8 lagging power factor. Since these are the rated values, this is expressed in normalized form as 1.0 pu volt-amperes and 1.0 pu voltage. Again taking the load voltage as reference phasor, we have

$$\hat{V}_2' = 1.0 + j0$$

$$\hat{I}_2' = 1.0\,(0.8 - j0.6)$$

$$\hat{V}_1 = 1.0 + j0 + 1.0 \times (0.8 - j0.6) \times (0.032 + j0.0752)$$

$$= 1.071 + j0.041$$

$$= 1.0715 \underline{/2.19}$$

Multiplying 25 000 V by 1.0715 gives 26 788 V, as before, for the primary voltage required to deliver this load.

The voltage regulation is now given directly by

$$\frac{1.0715 - 1.0}{1.0} = 0.0715 \text{ or } 7.15\%$$

The use of per-unit values is very common in the analysis of power systems, where the presence of many transformers makes it particularly valuable. While the advantages in the analysis of single transformers and machines are not as great, it does result in the numerical values in the calculations for all transformers being very similar to those found in the example above.

5.7 THREE-PHASE TRANSFORMERS

Transformers for use in three-phase systems may be formed of three separate single-phase transformers or they may be constructed on a common magnetic core normally having three limbs. The primary and secondary windings of each phase are wound concentrically on one limb. In all cases, the primary and secondary windings may be connected either in wye or delta, and there is no requirement that both be the same. In fact, there are some situations, such as unbalanced loads, where it is advantageous to have one winding connected in wye and the other in delta. Figure 5-10 shows a typical connection for such a transformer. In the case of the wye-connected side, the neutral will not be connected to a source if it is the primary, although it may be connected to ground. This also has the merit of minimizing the effect of harmonic currents.

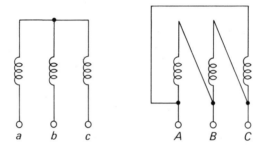

Figure 5-10   Three-phase Transformer with $Y = \Delta$ Connection.

At this stage it is usually convenient to refer all impedances to the wye-connected side of the transformer and work with the line-to-neutral equivalent as is common with the analysis of ordinary balanced three-phase circuits. If the other side is connected in delta, it is important to bear in mind during the entire calculations that the current is the phase current and thus must be multiplied by $\sqrt{3}$ when the line current is wanted.

Example 5-5

Three single-phase 2400/240-V transformers are connected to provide a three-phase transformer having a nominal voltage ratio 4160/240 V. Each has an equivalent resistance of 1.42 Ω referred to its HV winding and an equivalent reactance of 1.82 Ω also referred to its HV winding. Each transformer is rated 50 kVA.

Determine the primary voltage required and the corresponding voltage regulation when the transformer bank is delivering 150 kVA at 240 V and 0.8 lagging power factor.

Solution.   The 2400-V windings must be connected in wye if they are to be connected to a source of approximately 4160 V, and the 240-V windings must be connected in delta to the 240-V system. The load on each transformer is $150/3 = 50$ kVA and therefore the (phase) current flowing in the 240-V windings is given by

$$I_2 = \frac{50\ 000}{240} = 208.3\ \text{A}$$

However, if we are to refer all quantities to the wye-connected side, this current must be modified.

$$I'_2 = \frac{208.3 \times 240}{2400}$$

$$= 20.83\ \text{A}$$

Note that this could have been obtained directly as 50 000/2400 by recognizing that a voltage of 240 V when referred to the HV winding is 2400 V, and therefore the secondary current referred to the primary is $50\ 000/2400 = 20.83$ A. From this point the solution proceeds in the same manner as that of the corresponding single-phase problem.

$$\hat{V}'_2 = 2400 + j0$$

$$\hat{I}'_2 = 20.83(.8 - j0.6)$$

$$\hat{Z}_{eq} = 1.42 + j1.82 \text{ (referred to the 2400-V winding)}$$

$$\hat{V}_1 = 2400 + j0 + 20.83(0.8 - j0.6)(1.42 + j1.82)$$

$$= 2446.4 + j12.6 = 2446\ \underline{/0.3}$$

Since this winding is wye-connected, the (line) voltage required is 4237 V. The voltage regulation is

$$\frac{2446.4 - 2400}{2400} = 0.0193 \text{ or } 1.93\%$$

It is interesting to rework this problem using per-unit values. As with the single-phase example, the first step is to get the per-unit values of the equivalent resistance and reactance. Since the equivalent impedance is given referred to the HV winding, it is quicker to use rated current and voltage of this winding as the base values.

$$I_{unit} = \frac{50\ 000}{2400} = 20.83\ \text{A}$$

$$Z_{unit} = \frac{2400}{20.83} = 115.2\ \Omega$$

$$R_{pu} = \frac{1.42}{115.2} = 0.0123$$

$$X_{pu} = \frac{1.82}{115.2} = 0.0158$$

The load is the rated value at rated voltage, so the current is 1.0 pu at the power factor of 0.8 (lagging), and the expression for primary voltage is

$$\hat{V}_1 = 1.0 + j0 + 1.0 \times (0.8 - j0.6) \times (0.0123 + j0.0158)$$
$$= 1.01932 + j0.00526$$
$$= 1.01933 \underline{/0.29}$$

In a balanced three-phase system, any time the line-to-neutral voltage is 1.019 pu, the line-to-line voltage must also be 1.019 pu. The primary voltage is therefore $4160 \times 1.019 = 4239$ V. The small discrepancy is due to the fact that 4160 is slightly greater than $\sqrt{3}$ times 2400. When this value is used, the agreement is exact. The value of 1.93% for the voltage regulation, which may be seen directly from the per-unit primary voltage, is in exact agreement with that obtained previously.

5.8 AUTOTRANSFORMERS

Although most power transformers have two separate windings, it is possible to obtain similar transformation of voltage level using only one coil with a tapping point. This is shown schematically in Fig. 5-11, where N_1 represents the total number of turns, corresponding to the primary winding, and N_2 represents the turns in that part of the winding from which the output is taken, corresponding to the secondary. This is shown as a step-down arrangement, but the principle can be used equally well as a step-up transformer, since whatever is the set of primary turns, the magnetic field is produced and links with the entire winding. The currents are shown for a load connected across the section of winding having N_2 turns and the complete winding of N_1 turns connected to a source, V_1. The current in the section of winding that is common to primary and secondary is given by

$$\hat{I}_c = \hat{I}_2 - \hat{I}_1 \tag{5-32}$$

If the exciting current is negligible, the currents $\hat{I}_1$ and $\hat{I}_2$ are in phase, and I_c is simply the difference between the magnitudes of the two currents.

   In a single coil, the product of the cross section of each turn and the number of turns represents the minimum area required to accommodate it. Normally, the area actually required will be larger because of some space that cannot be occupied, especially with conductors having a circular cross

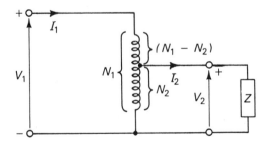

Figure 5-11   Autotransformer.

section. If the current density in the coil is J, the minimum area required for a coil is given by

$$A = \frac{NI}{J}$$

In a normal two-winding transformer, if the current density is the same in primary and secondary, the total cross section of copper required for the two windings is

$$A_2 = \frac{(N_1 I_1 + N_2 I_2)}{J}$$

For an autotransformer, the fact that I_c is the difference between input and output currents means that the corresponding area is

$$A_a = \frac{N_2 I_c + (N_1 - N_2)I_1}{J}$$

$$= \frac{N_2(I_2 - I_1) + (N_1 - N_2)I_1}{J}$$

and the ratio of the copper required is

$$\frac{A_a}{A_2} = \frac{N_2(I_2 - I_1) + (N_1 - N_2)I_1}{N_1 I_1 + N_2 I_2}$$

By noting that $N_1 I_1 = N_2 I_2$, this can be simplified to give the ratio as

$$\frac{A_a}{A_2} = 1 - \frac{N_2}{N_1} \qquad (5\text{-}33)$$

In addition to the reduction in the quantity of copper indicated by Eq. (5-33), there is also a reduction in the size of the core due to the smaller space required for the copper conductors.

Depending on the voltage ratio, an autotransformer may be significantly smaller than the corresponding two-winding transformer having the same rating. As a result, autotransformers are frequently used for very large units where there can be considerable difficulty in transportation from the manufacturer's plant to the site. The other common application is to variable-ratio transformers. These are usually wound on a toroidal core and a sliding contact used for connection to the secondary circuit. The main disadvantage is that there is no longer any isolation between the primary and secondary.

PROBLEMS

5-1. A single-phase, 60-Hz, 500-kVA transformer is rated 11 000/460 V. Its equivalent impedance, referred to the 11 000-V winding, has 5-Ω resistance and 20-Ω leakage reactance.

The transformer is to supply an inductive load of 500 kVA, 480 kW at 460 V.

(a) Find the primary voltage required.

(b) Determine the voltage regulation for the load condition given in part (a).

5-2. A certain single-phase, 11 000/550-V, 60-Hz, 100-kVA transformer has core losses of 500 W when operated at rated voltage. Its equivalent impedance, referred to the 11 000-V winding, has 25-Ω resistance and 75-Ω leakage reactance.

The transformer is to supply 20 kVA at 550 V and a power factor of 0.9 (lagging).

(a) Find the primary voltage required.

(b) Determine the voltage regulation for the load condition given in part (a).

(c) Determine the efficiency for this load.

5-3. A single-phase, 440/220-V, 60-Hz, 25-kVA transformer is subjected to standard open circuit and short circuit tests. The measured values are:

Open circuit: 220 V, 10 A, 600 W (LV winding excited)
Short circuit: 35 V, rated current, 950 W (HV winding excited)

(a) Determine the impedances of the approximate equivalent circuit referred to the HV winding.

(b) Determine the impedances of the approximate equivalent circuit referred to the LV winding.

5-4. A transformer rated 230/110 V, 60 Hz, 25 kVA, is tested as follows. A no-load test performed on the 230-V winding gives 230 V, 10 A, 600 W. A short circuit test performed on the 230-V winding gives 35 V at rated current, 900 W.

(a) Determine the equivalent resistance and reactance of the approximate equivalent circuit in terms of the 110-V winding.

(b) Determine the efficiency at 1/2 rated load with a power factor of 0.8 (leading).

(c) Determine the voltage regulation at rated load on the 110-V winding at a power factor of 0.8 (lagging).

5-5. A single-phase transformer is rated 200/100 V, 60 Hz, 5 kVA. The core loss at rated voltage is 80 W, and the copper loss at rated current is 120 W. Determine the efficiency when delivering rated volt-amperes at rated voltage and a power factor of 0.8 (lagging).

5-6. A single-phase transformer is rated 2200/110 V, 60 Hz, 1000 kVA. It has a no-load loss of 3000 W at rated voltage and a copper loss of 9000 W at rated current. Determine its efficiency when subjected to the following loads:

(a) rated current at a power factor of 0.95 (lagging),

(b) 500 kW at a power factor 0.95 (leading),

(c) 250 kW at a power factor of 0.95 (lagging),

(d) 1000 kVA at a power factor of 0.0 (lagging),

(e) 1000 kVA at unity power factor.

5-7. A single-phase, 60-Hz, 500-kVA transformer is rated 11 000/460 V. Its equivalent impedance has 0.02 pu resistance and 0.05 pu leakage reactance. The transformer is to supply an inductive load of 500 kVA, 480 kW at 460 V.

(a) Find the primary voltage required.

(b) Determine the voltage regulation for the load condition given in part (a).

5-8. The equivalent impedance of a certain single-phase, 2300/460-V, 60-Hz, 100-kVA transformer, referred to the 2300-V winding, consists of 1.0-Ω resistance and 2.5-Ω leakage reactance. The transformer is to supply 100 kVA at 460 V and a power factor of 0.95 (lagging).
(a) Find the voltage required.
(b) Determine the corresponding voltage regulation.

5-9. A single-phase, 60-Hz, 200-kVA, 6000/460-V transformer has the following parameter values in ohms at 60 Hz.

$$R_1 = 2.0 \qquad X_1 = 5.0$$
$$R_2 = 0.02 \qquad X_2 = 0.05$$
$$R_m = 20\ 000 \qquad X_m = 5000$$

If the transformer is to deliver 190 kW and 65 kvar to an inductive load at 460 V, determine the primary voltage required, the corresponding voltage regulation, and the efficiency.

5-10. A certain single-phase, 60-Hz, 1000-kVA, 11 000/550-V transformer has the following parameter values in ohms at 60 Hz.

$$R_1 = 1.0 \qquad X_1 = 3.0$$
$$R_2 = 0.003 \qquad X_2 = 0.01$$
$$R_m = 10\ 000 \qquad X_m = 2000$$

The transformer is required to deliver 900 kW at a power factor of 0.95 at 550 V. Determine the primary voltage required, the corresponding voltage regulation, and the efficiency if the load is
(a) inductive,
(b) capacitive.

5-11. A single-phase, 60-Hz, 100-kVA, 11 000/550-V transformer has the following parameter values in ohms at 60 Hz.

$$R_1 = 10.0 \qquad X_1 = 30.0$$
$$R_2 = 0.03 \qquad X_2 = 0.1$$
$$R_m = 100\ 000 \qquad X_m = 20\ 000$$

If the transformer is to deliver 100 kVA, 90 kW at 550 V, determine the primary voltage required, the corresponding voltage regulation, and the efficiency when the load is
(a) inductive,
(b) capacitive.

5-12. A certain single-phase, 60-Hz, 20-kVA, 460/110-V transformer has the following parameter values in ohms at 60 Hz.

$$R_1 = 0.15 \qquad X_1 = 0.6$$
$$R_2 = 0.01 \qquad X_2 = 0.025$$
$$R_m = 800 \qquad X_m = 400$$

The transformer is required to deliver 20 kVA to a load at 110 V. Determine

the primary voltage required, the corresponding voltage regulation, and the efficiency if the power factor of the load is
(a) 0.85 inductive,
(b) 0.85 capacitive.

5-13. A single-phase, 60-Hz, 20-kVA, 550/110-V transformer has the following parameter values in ohms at 60 Hz.

$$R_1 = 0.2 \qquad X_1 = 0.75$$
$$R_2 = 0.012 \qquad X_2 = 0.025$$
$$R_m = 1000 \qquad X_m = 450$$

The impedance of the load connected to the transformer is equivalent to a resistance of 0.53 Ω in series with an inductive reactance of 0.29 Ω. If the primary is connected to a 550-V source, determine the load voltage, the corresponding voltage regulation, and the efficiency.

5-14. A single-phase, 60-Hz, 10-kVA, 460/110-V transformer has the following parameter values in ohms at 60 Hz.

$$R_1 = 0.15 \qquad X_1 = 0.5$$
$$R_2 = 0.012 \qquad X_2 = 0.045$$
$$R_m = 1500 \qquad X_m = 1000$$

The impedance of the load connected to the transformer is equivalent to a resistance of 1.1 Ω in series with a capacitive reactance of 0.4 Ω. If the primary is connected to a 460-V source, determine the load voltage, the corresponding voltage regulation, and the efficiency.

5-15. A single-phase, 60-Hz, 500-kVA transformer is rated 11 000/550 V. The per-unit equivalent resistance and reactance are 0.02 and 0.04, respectively. When the transformer supplies an inductive load of 450 kW and 62.5 kvar at 550 V, determine the primary voltage required and the corresponding voltage regulation.

5-16. A single-phase, 60-Hz, 200-kVA, 6000/460-V transformer has the following parameter values in ohms at 60 Hz.

$$R_1 = 2.0 \qquad X_1 = 5.0$$
$$R_2 = 0.02 \qquad X_2 = 0.05$$
$$R_m = 20\,000 \qquad X_m = 5000$$

The transformer is to supply 190 kW at a lagging power factor of 0.95 when connected to a 6000-V primary source. Determine the resulting voltage at the secondary terminals, the corresponding voltage regulation, and the efficiency. (Hint: An iterative solution may be required.)

5-17. A single-phase transformer is rated 2200/110 V, 60 Hz, 1000 kVA. It has a no-load loss of 3000 W at rated voltage and a copper loss of 9000 W at rated current. It is subjected to the following daily load cycle.

Time (h)	pu current	Power factor
3	0.2	0.8 lag
2	0.5	0.9 lag
3	0.9	0.9 lead
1	1.3	0.9 lag
1	1.0	0.8 lag
2	0.7	0.9 lag
3	0.5	0.8 lag
9	0.3	0.7 lag

Determine the energy efficiency for this load cycle.

5-18. Three single-phase, 550/120-V, 60-Hz, 10-kVA transformers are to be connected to transform from 550 V, three-phase to 208 V, three-phase. Each has equivalent resistance of 1.0 Ω referred to the HV winding, and equivalent reactance of 2.5 Ω also referred to the HV winding. The transformer bank delivers 30 kVA at 208 V at a power factor of 0.85 (lagging).

   (a) Draw a diagram showing clearly whether each winding is connected in wye or in delta.

   (b) Neglecting the magnetizing current in each transformer, determine the current flowing in each winding.

   (c) Determine the primary voltage required.

   (d) Determine the voltage regulation.

   (e) Determine the efficiency.

5-19. Three single-phase, 2300/460-V, 60-Hz, 100-kVA transformers are connected to form a three-phase, 4000/460-V transformer bank. The equivalent impedance of each transformer, referred to its HV winding, consists of 1.2-Ω resistance and 4.5-Ω reactance. The load on the transformer bank is 300 kVA, 120 kvar (inductive) at 460V.

   (a) Draw a circuit diagram showing whether each winding is connected in wye or delta.

   (b) Determine the current flowing in each winding, neglecting the magnetizing current.

   (c) Determine the primary voltage, the corresponding voltage regulation, and the efficiency.

5-20. Three single-phase, 460/208-V, 60-Hz, 20-kVA transformers are connected to form a three-phase, 460/208-V transformer bank. The equivalent impedance of each transformer, referred to its HV winding, consists of 0.4-Ω resistance and 0.9-Ω reactance. The load on the transformer bank is 60 kVA, 50 kW (lagging) at 208V.

   (a) Draw a circuit diagram showing whether each winding is connected in wye or delta.

   (b) Determine the current flowing in each winding, neglecting the magnetizing current.

(c) Determine the primary voltage, the corresponding voltage regulation, and the efficiency.

5-21. Three single-phase, 317/120-V, 60-Hz, 25-kVA transformers are connected to form a three-phase, 550/208-V transformer bank. The equivalent impedance of each transformer, referred to its HV winding, consists of 0.15-Ω resistance and 0.35-Ω reactance. The load on the transformer bank is 70 kW, 27 kvar (lagging) at 208V.

(a) Draw a circuit diagram showing whether each winding is connected in wye or delta.

(b) Determine the current flowing in each winding, neglecting the magnetizing current.

(c) Determine the primary voltage, the corresponding voltage regulation, and the efficiency.

6

DC MACHINES

6.1 INTRODUCTION

The dc motor has long been noted for the relative ease with which its speed can be changed. It has therefore been particularly suited for drive systems where the speed must be varied in a controlled manner. Its armature has the most complicated structure of the common industrial machines and, as a result, it is normally the most expensive. Despite this, in small sizes it is produced on a very large scale and used extensively in portable power tools with either ac or dc excitation. The smallest dc motors have power ratings of the order of one watt and the largest have ratings in excess of several megawatts. In its simplest form there are only two windings. The field winding consists of simple coils, one wound on each pole piece of the stationary frame and normally connected in series. The armature winding usually appears as a rather complicated set of coils in which the motional voltages are induced and in which the energy conversion takes place. For applications of the machine it is not necessary to know the details of armature windings. It is usually sufficient to know the form and orientation of the magnetic field produced.

In this chapter we shall consider the main points of dc machines, describing them only as much as is required to understand the basis of the ordinary model. The analysis will be limited to steady-state conditions so that the operating characteristics may be determined and appreciated. Chapter 9 includes an introduction to dynamic behavior.

One part of the dc machine which is the easiest to recognize is the commutator, since it is made of copper bars that are often visible. Perhaps the simplest way to describe its function is to consider it to be a mechanical rectifier. While this is true from the circuit point of view, it does not provide a complete picture, especially its effect on the magnetic field produced by the current flowing in the coils of the armature winding. Naturally, the designer of these machines must be familiar with all aspects, but normally the details are not essential for applications to the design of motor-drive systems.

6.2 BASIC STRUCTURE OF DC MACHINES

It is perhaps simplest to start by considering the voltage induced in a coil that is rotating in a uniform magnetic field. The voltage induced will not be constant, but will vary depending on the component of the velocity of each side of the coil, which is at right angles to the field at any instant. For this theoretical situation the induced voltage would be sinusoidal. As a practical detail we must note that access to the terminals of the coil must be by sliding contacts at a pair of *slip rings,* as shown in Fig. 6-1a. If each end of the coil is connected to a separate ring, two fixed terminals—called *brushes*—which are kept in contact with the rings will provide this access. A set of three slip rings used in a three-phase induction motor is shown in Fig. 6-2 to indicate the construction. The voltage between the pair of slip rings could be described by

$$e(t) = E_m \sin \omega t \qquad (6\text{-}1)$$

The waveform at the slip rings is shown in Fig. 6-1b, which clearly is not direct current; the next step is to consider how this alternating voltage may be made unidirectional. This is illustrated in Fig. 6-3a, where the two separate slip rings have been replaced by a single ring, which has been split into two half rings that are insulated from each other. Each end of the coil is now connected to one segment of the ring. The brushes remain in the same position as the coil rotates, so that each fixed terminal is always connected to the side of the coil where the relative motion between coil side and field is the same. The polarity of the voltage difference between the two fixed contacts is therefore always the same. That is, the voltage is now unidirectional, as shown in Fig. 6-3b.

If we now consider that a set of similar coils is distributed symmetrically and connected to a ring which has been split into a number of segments equal to the number of coil ends, we have a picture of the basic dc machine. In practice, the "split ring" is made of copper bars and is normally called a *commutator.* Figure 6-4 shows part of the commutator of a large dc motor, with some of the brushes in position. Because the bars of the

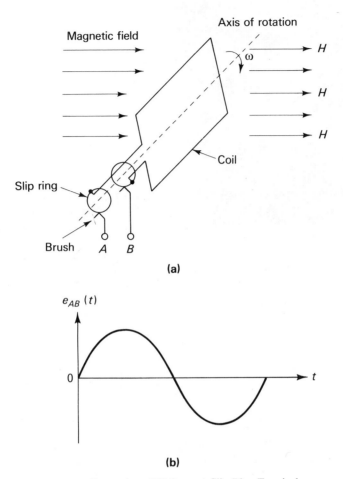

Figure 6-1   Generation of Voltage at Slip Ring Terminals.

commutator are held together by insulating material, it is not a very strong structure; this limits the maximum speed at which a dc machine may be operated.

   The commutator thus acts as a rectifier as far as the external circuit is concerned. It also has the effect that when any conductor is situated in a particular position *a*, as shown in Fig. 6-5, the current is always flowing in the same direction at that position. Underneath the pole face, as one conductor moves on from such a position, it is replaced by another carrying the same current. Thus, the magnetic field produced by the armature winding is stationary although the individual conductors are moving, and there is no relative motion between the component of magnetic field produced by the armature and that by the main field winding. In addition, the coils are

Figure 6-2 Set of Three Slip Rings (Photo courtesy of Canadian General Electric Co.).

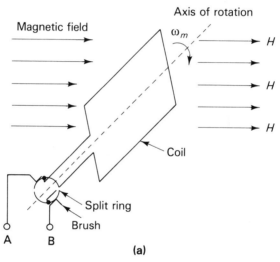

(a)

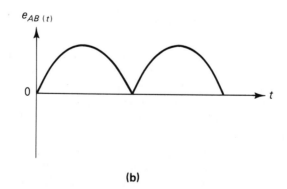

(b)

Figure 6-3 Generation of Voltage at Split Ring Terminals.

Figure 6-4 Commutator of DC Machine
(Photo courtesy of Canadian General
Electric Co.).

connected in series, so that at any instant most of them are contributing
to the induced voltage.

   In a real dc machine it is possible to arrange that each side of each
coil moves at right angles to a uniform magnetic field for about 70% of
each revolution. This is illustrated in the cross section shown in Fig. 6-6.
The pole structure is recognizable in that the coils are wound on cores that
project inward from the frame. The direction of the current is arranged to
produce alternate north and south poles, using the terminology of basic
physics. The simplest structure has only two such poles, but it is possible

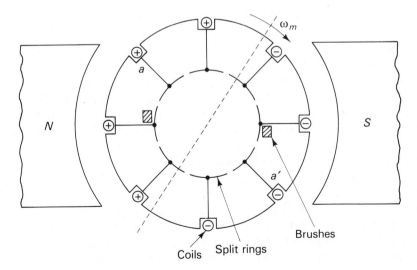

Figure 6-5 Current Distribution around Armature.

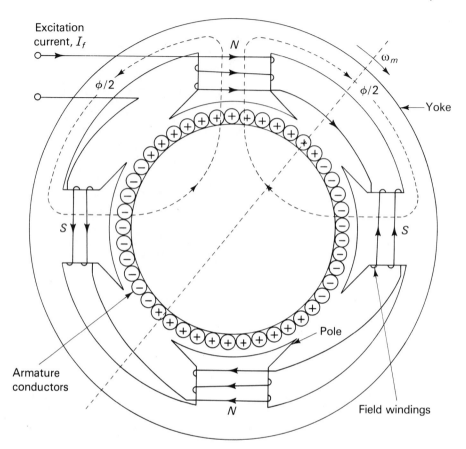

Figure 6-6 Cross Section of DC Machine.

to arrange additional pairs of poles so that there are several periods of field distribution around the air gap. This can be seen in Fig. 6-7, which is a view of the magnet frame of a large dc motor. It is then convenient to consider two scales by which angles are measured. The actual angle is usually expressed in *mechanical degrees,* and the angle within each period of field distribution is then expressed in *electrical degrees.* For example, in a four-pole machine each revolution of the shaft corresponds to 360 mechanical degrees or 720 electrical degrees.

Figure 6-8 shows a typical plot of the distribution of the radial component of flux density in the air gap of a dc machine. Each period of this distribution corresponds to a pair of poles excited by the field winding. The induced voltage thus has a small ripple, which does not cause any problems in practice. To complete this description of the dc machine, some armatures of large motors are shown in Fig. 6-9; the complete motor is shown in Fig. 6-10.

Figure 6-7 Magnet Frame of DC Motor (Photo Courtesy of Canadian General Electric Co.).

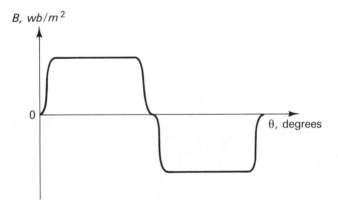

Figure 6-8 Field Distribution.

An expression for the voltage induced in an armature winding can now be obtained. The symbols used are:

B_a Average flux density in each pole, T
ϕ Flux per pole, Wb
P Number of poles (must be an even number)
l_a Axial length of the coil sides, m
D Diameter of the armature, m
Z Total number of conductors in the armature winding
a Number of parallel paths in the armature winding

E_a Average voltage, V

n Armature speed, revolutions per second

ω_m Angular velocity, rad/s

The voltage induced in a single conductor moving at velocity v at right angles to a uniform magnetic field is developed in most texts on electricity and magnetism. When applied to this situation it is given by

$$E_a = B_a l_a v \tag{6-2}$$

where

$$B_a = \frac{\phi}{l_a \dfrac{\pi D}{P}} = \frac{\phi P}{\pi l_a D} \tag{6-3}$$

and

$$v = \pi D n \tag{6-4}$$

Substituting in Eq. (6-2) gives

$$E_a = \frac{\phi P}{\pi D} \pi D n$$
$$= \phi P n \tag{6-5}$$

Each coil may have more than one turn and has two active sides. The total number of active conductors in the winding, Z, is thus given by the product of twice the number of coils and the number of turns per coil. The Z active conductors in an armature winding subdivide into a number

Figure 6-9   DC Rotors for Steel Mill Motors (Photo courtesy of Canadian General Electric Co.).

Figure 6-10   Large DC Motor for Steel Mill (Photo courtesy of Canadian General Electric Co.).

of parallel paths, a; this number depends on the details of the winding. There are therefore Z/a conductors connected in series in the winding. The voltage induced in the complete winding is therefore given by

$$E_a = \phi Z n \frac{P}{a} \tag{6-6}$$

$$= \frac{\phi Z \omega_m}{2\pi} \frac{P}{a} \tag{6-7}$$

There are two basic forms of armature winding in dc machines, *lap* and *wave*. These terms refer to the manner by which the individual coils are connected to each other. All that need be noted at this point is that a lap winding will always subdivide itself such that the number of parallel paths, a, equals the number of poles, P. A wave winding is divided such that the number of parallel paths is always two, irrespective of the number of poles. Figure 6-11 shows the general arrangement of a few coils for each type of winding.

The expression corresponding to Eq. (6-2) for the force that acts on a conductor in a uniform magnetic field is

$$f = B_a l_a i \tag{6-8}$$

where i is the current flowing through the conductor at right angles to the magnetic field. Since there are Z such conductors carrying a current I_a/a, the total force is

$$F = Z B_a l_a \frac{I_a}{a} \tag{6-9}$$

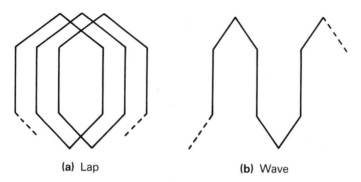

(a) Lap   **(b)** Wave

Figure 6-11   Lap and Wave windings.

The radius at which this force acts is taken as $D/2$, and the expression for the developed torque is

$$T_d = F\frac{D}{2} = ZB_a l_a \frac{I_a D}{2a}$$

Replacing the average flux density by $\phi P/(l_a \pi D)$ as before, we get

$$T_d = \frac{\phi Z I_a}{2\pi}\frac{P}{a} \tag{6-10}$$

The main purpose in developing these expressions is to bring out the basic properties of the dc machine. The induced voltage in a given machine is directly proportional to the flux per pole and to the speed. All other factors in these expressions are design constants which cannot be changed after a machine has been built.

An important point to note is that when the induced voltage is multiplied by the armature current, the result is an expression for the power being converted at the air gap of the machine. Using Eq. (6-7), this product is

$$E_a I_a = I_a \frac{\phi Z \omega_m}{2\pi}\frac{P}{a}$$

$$= \frac{I_a \phi Z}{2\pi}\frac{P}{a}\omega_m \tag{6-11}$$

$$= T_d \omega_m$$

which is the gross mechanical power in watts when the speed is in radians per second.

6.3 DC GENERATORS

Although dc generators generally find very little application, it is probably easier to apply the model of a dc machine to motoring problems after developing it by consideration of the dc generator. The main thing to keep

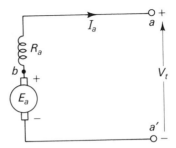

Figure 6-12 Steady-State Circuit Model
of Armature.

in mind when writing the voltage equation for the armature circuit is that
there are two concurrent phenomena taking place in the armature. First
there is the rise in voltage (E_a) due to induction when the coils are in motion
traversing the magnetic field. This is often called the *speed voltage*. There
is also a drop in voltage $(I_a R_a)$ due to the resistance of the winding, as in
any passive dc circuit. These are represented in Fig. 6-12 where the points
aa' are the actual terminals of the armature. The junction between the
symbol for the voltage source and that for the resistance of the armature,
b, has no physical counterpart. For dynamic problems the self-inductance
of the armature must also be included as a third element connected in series
in the circuit model.

The magnetic field is normally supplied by means of a set of coils
placed on each pole piece and known collectively as the *field winding,*
although there are examples of dc machines where the field is produced
by a set of permanent magnets. There are four main arrangements of
supplying current to the field winding:

(a) Separate excitation
(b) Shunt excitation
(c) Series excitation
(d) Compound excitation

6.3.1 Separate Excitation

With separate excitation as shown in Fig. 6-13, the field current is
independent of the terminal voltage at the armature, and thus any change
in armature current has no resultant change on the field current. If this
were a complete model of all that takes place in a dc machine, a constant
value of field current would produce a constant value of magnetic flux in
each pole, and the induced voltage (E_a) would be independent of armature
current. The terminal voltage is therefore given by the expression

$$V_t = E_a - R_a I_a \tag{6-12}$$

Unfortunately, the magnetic field produced by the armature current causes
a small reduction in the flux per pole, and E_a decreases slightly as I_a

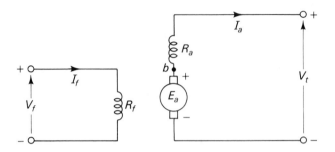

Figure 6-13 Separately Excited Generator.

increases. This action is called *armature reaction* and is considered later in Sect. 6.6.

6.3.2 Shunt Excitation

With shunt excitation the field winding is connected directly across the armature winding as shown in Fig. 6-14. In this case any change in armature current will cause a change in the resistive voltage drop $(I_a R_a)$. As a result, both the terminal voltage and the field current must also change, and thus the induced voltage (E_a) is now dependent on the armature current. However, Eq. (6-12) is still a valid expression for the terminal voltage at the armature; the difference in this case is that E_a is no longer constant, even when the effect of armature reaction is neglected.

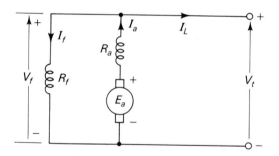

Figure 6-14 Shunt Excited Generator.

The terminal voltage with shunt excitation may be obtained using the *magnetization* or *open circuit characteristic* (*occ*). This is a plot of the speed voltage induced in the armature when it is rotated at some constant speed. A typical characteristic is shown in Fig. 6-15.

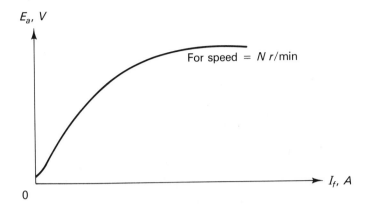

Figure 6-15   Open Circuit Characteristic.

Due to saturation of the magnetic core, the induced voltage also saturates, and eventually a large increase in field current is required to produce only a small increase in voltage. This characteristic is normally obtained by exciting the field coil separately, irrespective of the intended connection and irrespective of whether the machine is to operate as a generator or as a motor. Since the armature current is zero during these measurements, the terminal voltage is numerically equal to the speed voltage. This characteristic therefore gives the general relationship between speed voltage and field current when the armature rotates at this particular speed. To obtain the characteristic for any other speed it is a simple matter to note that Eqs. (6-6) and (6-7) are valid for all values of field current and may therefore be applied to any number of sample values as required. That is, for any value of field current the speed voltage is directly proportional to the speed, and the ordinates at another speed are obtained by multiplying the values at the first speed by the ratio of the speeds.

To obtain the load characteristic we must combine the open circuit characteristic with that of the field winding in the steady state. These are shown in Fig. 6-16a, where the characteristic of the field is simply that of an ohmic conductor, namely

$$V_t = R_f I_f \tag{6-13}$$

Normally the field current is of the order of 5% of the rated armature current and the $I_a R_a$ voltage drop is negligible since the only current flowing through the armature is the field current. The intersection of the two characteristics is the only point that can simultaneously satisfy both characteristics; it is therefore the operating point of the shunt generator on no-load.

To obtain the operating point when the armature current is large enough to cause a significant drop in terminal voltage [Eq. (6-12)], the same

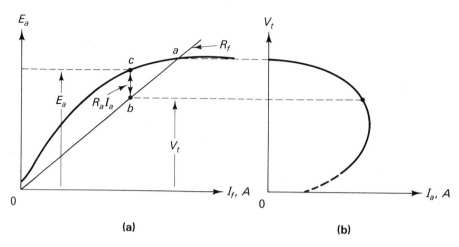

Figure 6-16 Open-Circuit and Load Characteristics of Shunt Generator.

two characteristics are used. In this case the only thing to note is that the operating point must lie on the field resistance line, since this represents the voltage-current relationship of the field winding under all steady-state conditions. If for the moment we assume that the operating point is b in Fig. 6-16a, the speed voltage for this field current is that at point c. The difference between these two values of voltage must be the armature resistive voltage drop, $I_a R_a$, at least if the armature reaction effect is neglected. Dividing this value of voltage drop by the armature resistance gives the armature current for this terminal voltage. This may be repeated with sufficient points to plot the load characteristic shown in Fig. 6-16b. If desired, the values of field current may be subtracted from these values of armature current so that the terminal voltage is plotted as a function of load current.

As with the transformer, voltage regulation is the difference in voltage between no-load and a load condition, divided by the load voltage. In the case of a self-excited shunt generator this is *not* equal to $I_a R_a$. The no-load voltage and the terminal voltage on load must be obtained using the procedure described above.

6.3.3 Series Excitation

In this case the armature and field windings are connected in series, as shown schematically in Fig. 6-17. The terminal voltage for series excitation is obtained in a manner similar to that shown in Sec. 6.3.2 for shunt excitation. The main difference is that the load characteristic is obtained directly from the magnetization characteristic by noting that field current and armature current are now constrained to be equal.

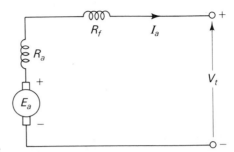

Figure 6-17   Series Excited Generator.

The load characteristic is plotted in Fig. 6-18 as the curve V_t. If the load resistance is known, its characteristic can be plotted in the same manner as that of the field resistance of the shunt generator. The intersection of this load line and the terminal voltage characteristic that has just been obtained, point b in Fig. 6-18, is the operating point. As in the case of the shunt generator, this solution assumes that the armature reaction effect is negligible.

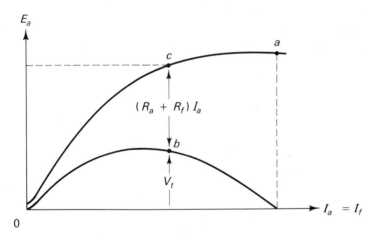

Figure 6-18   Series Generator Characteristics.

6.3.4 Compound Excitation

It is possible to modify the characteristics of a shunt generator by adding a set of coils to the field system to carry the load current. Figure 6-19 shows schematic diagrams of the connections. When the series field aids the shunt field, the arrangement is called *cumulative compound* excitation; when it opposes the shunt field it is *differential compound* excitation. Often such a winding is intended to compensate, at least partially, the effect of

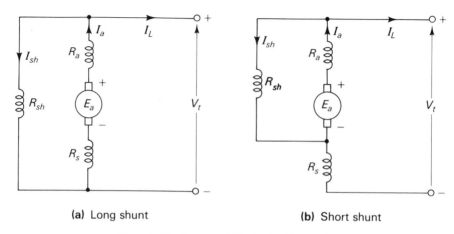

(a) Long shunt   **(b)** Short shunt

Figure 6-19 Compound Excitation Connections.

armature reaction; such cumulative compounding is used in both motors and generators. In the examples that follow, the field current is used directly. For compound excitation, the procedure is the same except that the total mmf on the magnetization axis must be used.

Example 6-1

A certain dc machine has the magnetization characteristic given below when its armature speed is 1500 r/min. The resistance of its field winding, R_f, is 50 Ω, and that of its armature, R_a, is 0.2 Ω.

I_f	0	1.0	2.0	3.0	4.0	5.0
E_a	5	94	166	200	215	227

It is to operate as a shunt generator with a suitable resistance, R, connected in series with the field winding. The value of R is to be chosen so that it will always be possible to adjust the terminal voltage to 200 V for the following conditions of load current and shaft speed. Find the value of R, assuming negligible armature reaction, when

(a) the speed is 1500 r/min and $I_a = 0$,
(b) the speed is 1500 r/min and $I_a = 20$ A,
(c) the speed is 1450 r/min and $I_a = 20$ A.

Find the voltage regulation for the load conditions of parts (b) and (c), assuming that the speed is 1500 r/min when there is no load on the generator.

Solution

   (a) The magnetization characteristic at 1500 r/min is plotted and shown in Fig. 6-20. The resistance R and the field winding form a series connection, and when this characteristic is plotted it must intersect the magnetization characteristic

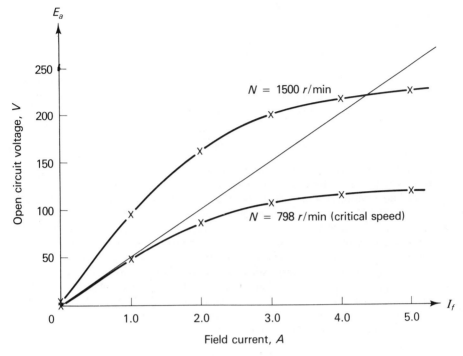

Figure 6-20   Magnetization Characteristic, Example 6-1.

at 200 V if this is to be the open circuit voltage as required in part (a). Since the field current required is 3.0 A, the equivalent resistance of the field current is

$$R_f + R = \frac{200}{3} = 66.67 \ \Omega$$

and

$$R = 66.67 - 50 = 16.67 \ \Omega$$

(b) The value 200 V must be a point on the field circuit resistance line such that the difference between the terminal voltage and the speed voltage at this (still unknown) value of field current is

$$I_a R_a = 20 \times 0.2 = 4.0 \text{ V}$$

Clearly the speed voltage must be greater than that in part (a). When the terminal voltage is 200 V, the speed voltage must be 204 V, which must be a point on the magnetization characteristic at 1500 r/min. Linear interpolation gives the required field current as

$$I_f = 3.27 \text{ A}$$

and the equivalent resistance of the field circuit is obtained by dividing the terminal voltage by this value of field current:

$$R_f + R = \frac{200}{3.27} = 61.2 \ \Omega$$

Thus

$$R = 11.2 \ \Omega$$

To obtain the voltage regulation we must first obtain the open circuit voltage when the generator is driven at a speed of 1500 r/min and the total resistance in the field circuit is 61.2 Ω. This is obtained from the intersection of the open circuit characteristic at 1500 r/min and the field resistance line and is found to be approximately at 205 V. The voltage regulation is therefore

$$\text{Regulation} = \frac{205 - 200}{200} = 2.5\%$$

Note that the speed voltage (204 V) does not equal the open circuit voltage because the field is reduced as the load current is increased.

(c) The solution to this part starts off as in part (b), and the speed voltage is obtained as 204 V, as before. However, in this case the value of 204 V is a point on the magnetization characteristic at 1450 r/min. If we wish, we can obtain this characteristic by multiplying all the voltage values given in the table by 1450/1500. This can be plotted, and the solution would then be the same as that in part (b). Fortunately, this is not necessary, since we can obtain the corresponding point on the 1500 r/min characteristic by noting that for the same value of field current a speed voltage of 204 V at 1450 r/min becomes 204 $\times$ 1500/1450 = 211 V at 1500 r/min. Linear interpolation at E_a = 211 on the 1500 r/min characteristic gives the necessary field current as 3.73 A. The solution now proceeds as before.

$$R_f + R = \frac{200}{3.73} = 53.6 \ \Omega$$

and

$$R = 3.6 \ \Omega$$

The voltage regulation is obtained as in part (b), noting that the open circuit voltage is obtained from the intersection of the 1500-r/min characteristic and the 53.6-Ω field resistance line. The value is found to be approximately 215 V with the result

$$\text{Regulation} = \frac{215 - 200}{200} = 7.5\%$$

If the field resistance can be controlled within the limits that have just been calculated, it is theoretically possible to reduce the effective voltage regulation to zero.

Note that if the speed drops below a critical value where the field resistance line is tangential to the magnetization characteristic, the speed voltage drops off very rapidly. In this example the *critical speed* can be estimated by finding the speed at which the first tabulated value other than I_f = 0 (94 V at I_f = 1 A) coincides with the voltage that would appear across the field winding ($R_f I_f$ = 50 $\times$ 1), namely 1500 $\times$ 50/94 = 798 r/min. Since the residual field normally causes the straight portion of the magnetization characteristic to be a line that does not quite pass through the origin, the critical speed cannot be stated with precision; in this case it is approximately 800 r/min.

6.4 DC MOTORS

Referring again to Eq. (6-12), it will be seen that if for any reason the induced voltage becomes less than the terminal voltage, the direction of the armature current must reverse and thus the developed torque must also reverse. As far as the analysis of dc machines is concerned, this is the essential difference between a motor and a generator. There are also differences in the design of a dc machine, depending on whether it is intended for motor operation or for generator operation. However, the circuit model is virtually the same. In the case of a motor, it is more convenient to show the positive flow of current going into the armature terminal where the voltage is positive. The speed voltage now opposes the flow of current; for this reason, it is sometimes called the *back emf*. This is shown in Fig. 6-21 and is the same convention as is used in a passive circuit where the elements cannot generate energy but can only absorb it.

The application of a dc motor to a particular situation will in general require consideration of five relationships. The first of these is the armature circuit voltage equation, and with the choice of polarity noted above, it becomes

$$V_t = E_a + R_a I_a \tag{6-14}$$

The second relationship to be considered is the developed or electromagnetic torque, which was shown to be proportional to both armature current and flux per pole [Eq. (6-10)]; for motor applications it is usually expressed as

$$T_d = K_t I_a \phi \tag{6-15}$$

where ϕ still represents the flux per pole.

The third relationship is the corresponding expression for the speed-induced voltage. For motor applications Eq. (6-7) is usually expressed as

$$E_a = K_v \phi \omega_m \tag{6-16}$$

and it may be shown that, provided the speed is expressed in radians per second, the constants K_t and K_v are equal.

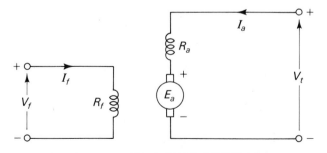

Figure 6-21   Circuit Model of DC Motor.

The fourth relationship is due to saturation of the core as expressed by the magnetization characteristic. It is used to give the effective relation between flux per pole and field current, although this is normally given in the form of a graph of voltage as a function of field current. This is the same magnetization characteristic as was described and used in the discussion of generator operation.

The fifth characteristic that must be considered in any motor application is that of the system being driven. The steady-state torque-speed characteristic may have constant torque, or torque proportional to speed, or the torque is some other function of speed. It will usually be necessary to consider this characteristic in order to determine the armature current.

The prediction of the performance of a dc motor requires that all the five relationships noted above be satisfied simultaneously. In effect, this requires the solution of five simultaneous equations where at least one is not available in functional form. Fortunately, all five variables do not appear in all the characteristics, and it is usually possible to obtain a solution by elimination of variables. This is best illustrated by the examples that follow. Before proceeding to these examples, let us first investigate the characteristics that result from the five relationships above.

If a dc motor is supplied from a constant voltage source, there is no difference between shunt excitation and separate excitation, since the field current is independent of load. With constant flux, Eq. (6-16) may be written in terms of the speed and Eq. (6-14) substituted for the speed voltage. The result is

$$\omega_m = \frac{1}{K_v \phi}(V_t - R_a I_a)$$
$$= \frac{1}{K_v \phi}\left(V_t - \frac{R_a}{K_t \phi}T_d\right)$$

$$(6\text{-}17)$$

If for the moment we neglect the mechanical losses in the motor, the load torque equals the developed torque, and Eq. (6-17) represents the load characteristic. The basic speed-torque characteristic of the dc shunt or separately excited motor is therefore a straight line which droops slightly as the load torque is increased. Normally the armature resistance is low, so that the slope of the characteristic also is low. The solid line shown in Fig. 6-22 represents this characteristic. We have already noted that in a typical machine the flux decreases slightly as the armature current increases. Examination of Eq. (6-17) shows that this will tend to increase the speed corresponding to any particular value of load torque, and the dashed line in Fig. 6-22 shows the resulting characteristic. Note particularly that with heavy loads, or also when there is a sudden increase in load, there is a distinct possibility that the motor will become unstable, as indicated by the rising characteristic.

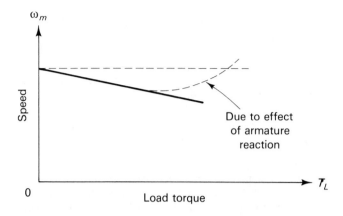

Figure 6-22   Load Characteristic of Shunt Motor.

Example 6-2

A certain dc shunt motor has the following magnetization characteristic at 1195 r/min.

E_a	230	240	250	260	270
I_f	1.05	1.13	1.26	1.46	1.67

Its armature resistance is 0.25 Ω and its field resistance is 198 Ω. Determine the operating speed for the following conditions. The effect of armature reaction is to be neglected.

(a) The motor is connected to a 250-V source on no load.

(b) The terminal voltage is changed to 230 V with no load.

(c) The terminal voltage is changed to 270 V with no load.

(d) The terminal voltage is 250 V and the load torque is such that the armature current is 30 A.

Solution. If there is no (mechanical) load placed on the shaft, the torque required of the motor is only that required to overcome friction in its bearings and *windage,* which is the term used for the torque required by the motion of the armature in air. Windage may also be increased by the necessity of a fan to provide cooling of the armature and field. Ideally, both perhaps should be zero, in which case the developed torque required would be zero and, from Eq. (6-15), the armature current would also be zero. In practice there is a small current, but this is normally so small that the resistive voltage drop ($I_a R_a$) is very much less than the terminal voltage, and the induced voltage is almost the same as the terminal voltage. Thus, when a motor has no load on its shaft, it is usually justifiable to take the induced voltage as equal to the terminal voltage.

(a) With the field winding connected directly to the 250-V source, the field current is given by

$$I_f = \frac{250}{198} = 1.26 \text{ A}$$

From the magnetization characteristic we may observe that when the speed is 1195 r/min, and the field current is 1.26 A, the induced voltage is 250 V. Since the actual value of induced voltage is 250 V and the actual field current is 1.26 A, then the operating speed must be that of the magnetization characteristic, namely 1195 r/min.

(b) With 230 V across the field winding, the field current becomes

$$I_f = \frac{230}{198} = 1.16 \text{ A}$$

Again referring to the magnetization characteristic, we may observe that when the field current is 1.16 A and the speed is 1195 r/min, the induced voltage is 242.3 V (obtained by linear interpolation). But the actual value of the induced voltage is 230 V, since there is no load, and thus the motor must slow down until the speed voltage or back emf becomes 230 V. Since this implied change takes place at constant field current, the speed is directly proportional to the induced voltage and the resulting no-load speed is

$$N_o = 1195 \times \frac{230}{242.3} = 1134 \text{ r/min}$$

(c) In this case the increase in terminal voltage may reasonably be expected to cause an increase in speed. The argument is the same as that in part (b). The field current is

$$I_f = \frac{270}{198} = 1.36 \text{ A}$$

The magnetization characteristic shows that when the field current is 1.36 A and the speed is 1195 r/min, the induced voltage is 255 V, again obtained by linear interpolation. With this value of field current the armature must speed up until the induced voltage equals 255 V, and so the no-load speed is

$$N_o = 1195 \times \frac{270}{255} = 1265 \text{ r/min}$$

(d) The motor is now loaded such that the armature current is 30 A. The more general situation where the induced voltage and the terminal voltage are not equal now prevails. The induced voltage is

$$E_a = V_t - I_a R_a$$
$$= 250 - 30 \times 0.25 = 242.5 \text{ V}$$

Since the terminal voltage is 250 V, as in part (a), the field current is again 1.26 A. Referring to the magnetization characteristic, we note that when the field current is 1.26 A and the speed is 1195 r/min, the induced voltage is 250 V rather than the actual value of 242.5 V. Similarly to part (b), the armature must slow down until the speed voltage (back emf) is 242.5 V. That is,

$$N = 1195 \times \frac{242.5}{250} = 1159 \text{ r/min}$$

In effect, as torque is demanded from the motor, it is developed by the motor reducing its speed until the armature current is sufficient to develop the required torque.

Example 6-3

A certain dc shunt motor has an armature resistance of 0.05 Ω. It drives a mechanical system whose torque is proportional to speed at a speed of 700 r/min with an armature current of 200 A when connected to a 500-V source. A simple, although inefficient, way of reducing the speed is to insert resistance in series with the armature so that the armature circuit resistance is increased. Determine the value of the additional resistance required to reduce the speed to 600 r/min. The effect of armature reaction is to be neglected.

Solution. Since the terminal voltage is presumed to be constant and the field winding is connected directly to the source, the field current will not be affected by any change in the armature circuit. Initially, the speed voltage is

$$E_{a1} = 500 - 200 \times 0.05 = 490 \text{ V}$$

Finally, this speed voltage must be changed so that with the same field, the speed is proportional to it. Thus, the speed voltage required is

$$E_{a2} = 490 \times \frac{600}{700} = 420 \text{ V}$$

and this value must satisfy the voltage equation

$$E_{a2} = 500 - I_a(R_a + R)$$

where neither I_a nor R is known. This is the point where we must consider the properties of the mechanical system. Its characteristic was stated as "torque proportional to speed." That is, if the friction and windage of the motor are negligible, the developed torque must also be proportional to speed. Referring to the torque equation,

$$T_d = K_t \phi I_a$$

we can note that if the flux does not change and T_d is proportional to speed, the armature current I_a must also be proportional to speed. That is, at 600 r/min the armature current will be

$$I_a = 200 \times \frac{600}{700} = 171.4 \text{ A}$$

and the voltage equation becomes

$$420 = 500 - 171.4 \times (0.05 + R)$$

from which the additional resistance is found to be 0.417 Ω.

Current practice is to use power electronics circuits to control the armature voltage with high efficiency. However, the principle remains the same. When the field is kept constant, the speed, being directly proportional to the induced voltage, is almost but not exactly proportional to the armature voltage. If the mechanical system characteristic is that of constant torque, the problem above is simpler in that the steady-state armature current is independent of speed and the value of resistance required to reduce the speed to 600 r/min becomes 0.35 Ω.

Example 6-4

A certain mechanical system is to be driven at 2000 r/min by a dc motor. It is necessary to use an existing dc shunt motor which has the following values.

$$R_a = 0.3 \ \Omega \qquad R_f = 200 \ \Omega$$

The rated voltage is 500 V, and it has been determined that when armature and field are both connected to a 500 V source, the motor is capable of driving the system at a speed of 1746 r/min, requiring an armature current of 50 A. The load torque is independent of speed.

It is proposed to adapt the motor for this application by weakening the field; its magnetization characteristic at 1800 r/min is as follows.

E_a	300	386	450	500	540	573
I_f	1.0	1.5	2.0	2.5	3.0	3.5

Determine the value of resistance to be placed in series with the shunt field so that the motor may drive the mechanical system at 2000 r/min. Armature reaction is to be neglected.

Solution. Although there is a constant torque load, the fact that the speed is to be changed by reduction of the field current means that the armature current must increase [Eq. (6-15)]. The solution requires considerable use of the magnetization characteristic which should at least be sketched, if not plotted accurately. Figure 6-23 shows such a sketch. The first step, as in the previous example, is to obtain the value of the speed voltage at the known operating point.

$$E_a = 500 - 50 \times 0.3 = 485 \ \text{V}$$

This must be a point on the magnetization characteristic at 1746 r/min, but it is more convenient to determine the corresponding value at 1800 r/min so that the given characteristic may be used directly. This is

$$E_a' = 485 \times \frac{1800}{1746} = 500 \ \text{V}$$

As before, the magnetization characteristic is the effective relationship between flux and field current. If the speed voltage E_a' were to remain 500 V, we would be faced with a variation of the problem in Example 6-1. That is, we can use the magnetization characteristic to obtain the field current that would result in a speed of 2000 r/min when the speed voltage is 500 V. If we imagine that the characteristic were redrawn for 2000 r/min, the point 500 V must lie on it at the value of field current required. But the corresponding point on the 1800 r/min characteristic is

$$E_{1800} = 500 \times \frac{1800}{2000} = 450 \ \text{V}$$

and the field current is found to be 2.0 A.

The magnetization characteristic can now be used to get the ratio of the flux per pole for the two conditions, since the voltages on a magnetization characteristic are directly proportional to the flux. Referring now to Eq. (6-15), we can see that when the torque is constant, the armature currents for the two conditions must

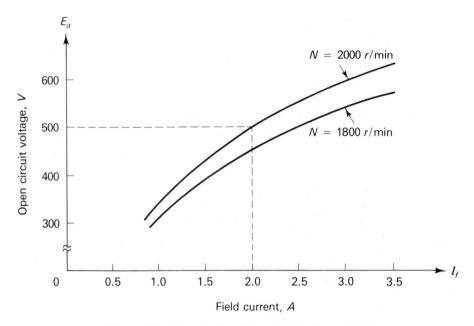

Figure 6-23   Magnetization Characteristic, Example 6-4.

have the inverse ratio of the fluxes. Since we must use voltages from the magnetization characteristic as proportional to the corresponding fluxes, the armature current is given by

$$I_a = 50 \times \frac{500}{450} = 55.6 \text{ A}$$

and the speed voltage is

$$E_a = 500 - 55.6 \times 0.3 = 483.3 \text{ V}$$

This value is less than the 485 V estimated at the beginning of the solution, so these calculations can be repeated until the solution converges, which it usually does rapidly. The results of the second iteration are

$$E_{1800} = 435 \text{ V}$$

$$I_f = 1.88 \text{ A}$$

$$I_a = 57.5 \text{ A}$$

$$E_a = 482.8 \text{ V}$$

Since this is very close to the value of 483.3 V above, it may reasonably be taken as the solution, and the additional field resistance is found from

$$R_f + R = \frac{500}{1.88} = 266 \text{ } \Omega$$

$$R = 66 \text{ } \Omega$$

In all the examples so far, the field winding has been connected directly across the terminals of the source in a shunt connection. However, there are many examples, especially in electric traction, where the field winding consists of a few turns of conductors capable of carrying the full armature current. The series motor has a torque-speed characteristic which is better suited to the traction application at low speeds when high accelerating torque is usually required. However, with the development of control systems based on power electronics, separately excited motors are also used in traction.

Series motors in which the entire field structure is laminated operate satisfactorily on ac as well as on dc. These are the universal motors which are used to drive most portable tools in industry and at home. They will not be considered further in this text.

As far as the analysis of the dc series motor is concerned, the main point to note is the constraint that armature and field currents are equal. At light load, since the armature current decreases, so also does the field, with the result that the speed increases very rapidly. Care must be taken to ensure that the load cannot be removed completely. This characteristic can be obtained by considering that the flux is proportional to the field current at low values.

$$\phi = K_a I_a \tag{6-18}$$

The developed torque is therefore given by

$$T_d = K_t K_a I_a^2 \tag{6-19}$$

Thus as long as the core is not saturated, the torque developed by a series motor is proportional to the square of the current. To obtain the speed-torque characteristic, Eq. (6-17) can be rewritten as

$$\omega_m = \frac{1}{K_v K_a I_a}\left(V_t - \frac{R_a}{K_t K_a I_a}T_d\right) \tag{6-20}$$

$$= \left[\frac{V_t\sqrt{K_t}}{K_v\sqrt{K_a}}\right]\frac{1}{\sqrt{T_d}} - \frac{R_a K_a}{K_v K_a^2}$$

This inverse relation with the developed torque is shown in Fig. 6-24; any application of the series motor must ensure that there is no possibility of a complete removal of the load torque. This problem of the series motor on light load can be partly resolved by adding a shunt cumulative winding. This ensures that there is always at least some minimum field present.

Example 6-5

A dc series motor operates at 450 r/min with a current of 80 A when connected to a 600-V source. The resistance of its armature and field windings is 0.45 Ω, and the flux may be taken as proportional to current.

If the torque is changed to half of that above, determine the current and the speed at which it will operate, neglecting the effect of armature reaction.

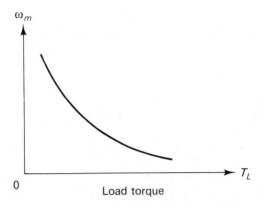

ω_m

0

Load torque

T_L

Figure 6-24   Load Characteristic of Series Motor.

Solution.   Although Eqs. (6-18) to (6-20) are applicable, it is perhaps more instructive to return to Eqs. (6-14) to (6-16) to solve this problem. The initial speed voltage is

$$E_{a1} = 600 - 80 \times 0.45 = 564 \text{ V}$$

Before we can determine the final value of the speed voltage, we must first determine the armature current. Since the flux is proportional to current, Eq. (6-15) shows that the developed torque is proportional to the square of the armature current. Thus the current becomes

$$I_{a2} = 80 \times 0.5^2 = 20 \text{ A}$$

so that

$$E_{a2} = 600 - 20 \times 0.45 = 591 \text{ V}$$

and Eq. (6-16) can be applied, recognizing that the ratio of the values of flux for the two operating conditions is given by the ratio of the currents. That is,

$$\frac{\omega_{m2}}{\omega_{m1}} = \frac{E_{a2}\phi_1}{E_{a1}\phi_2} = \frac{E_{a2}I_1}{E_{a1}I_2} = \frac{591 \times 80}{564 \times 20} = 4.19$$

and the new operating speed is $4.19 \times 450 = 1886$ r/min.

6.5 EFFICIENCY

From an external point of view, a dc machine can be regarded as a simple dc device. However, it should now be evident that internally, the current and voltage in the individual conductors are alternating, although they are not sinusoidal. The fact that the core of the armature is subjected to repeated flux reversals means that there are hysteresis and eddy current losses just as in a transformer. As a result, it is necessary to laminate the armature core in order to reduce the core losses. Since these exist only when the armature is rotating, they are *rotational losses* which are fed from the shaft. In addition, there are the I^2R losses in the armature and in the field. The ordinary mechanical losses consist of friction in the bearings and at the commutator as well as windage. These losses are present in every dc machine, whether it is a motor or a generator. The efficiency is simply the ratio of the actual output power to the input power. When the output power is known, the input power is obtained by adding all the rotational and electrical losses. The power required to excite the field must be included among the losses.

The rated value of output power is generally stated on the nameplate of the machine, which also includes rated terminal voltage and excitation. In the case of a motor it will also include the *base speed*—that is, the speed at which it will operate with rated armature voltage and field current— and the maximum permissible speed from the point of view of mechanical safety. Thus, to operate a motor at speeds higher than base speed, but at rated voltage, it is necessary to weaken the field.

Example 6-6

When a certain dc shunt motor is connected to a 240-V source and allowed to operate with no external load connected to its shaft, the armature current is 25 A, the field current is 15 A, and the speed is 1100 r/min. Its armature resistance is 0.03 Ω. When the load is applied to the shaft, the armature current increases to 400 A and the speed drops to 1050 r/min. Determine the corresponding efficiency.

Solution. It is implied that the terminal voltage is constant, so that the power required to excite the field is

$$P_f = 240 \times 15 = 3600 \text{ W}$$

The rotational losses on no-load are

$$P_o = 25(240 - 25 \times 0.03) = 5981 \text{ W}$$

These losses comprise the core losses, windage, and friction. Each of these is a different function of speed, and as long as the change in speed is minimal, they may be considered to be constant. Since the drop in speed on applying the load torque is very small in this case, the rotational losses may reasonably be taken as 5981 W.

When the motor is loaded, the output power is given by

$$P_{\text{out}} = 400(240 - 400 \times 0.03) - 5981 = 91\,200 - 5981 = 85\,219 \text{ W}$$

and the input power is

$$P_{\text{in}} = 240 \times 400 + 3600 = 99\,600 \text{ W}$$

The efficiency is thus $85\,219/99\,600 = 85.6\%$.

6.6 ARMATURE REACTION

In the discussion so far we have considered only the magnetic field produced by the field winding, this being the predominant effect as far as the simple circuit model is concerned. However, the armature, as was noted before, is an interconnection of a number of coils, and it is inevitable that the armature current will produce a magnetic field. Indeed, if this were not so, there would be no conversion of energy.

It can be shown that the magnetic field produced by the armature is distributed in space, although not in the same manner as that due to the field winding as shown in Fig. 6-8. Figure 6-25b shows the mmf distribution for the idealized situation where the field winding produces a perfectly rectangular distribution of mmf. Because the armature winding is uniformly distributed around the surface of the armature, a plot of current as a function of position is also a rectangular waveform as shown in Fig. 6-25c. When Eq. (3-4) is applied to this situation, the mmf is given by

$$\mathscr{F}_a(\theta) = \int H d\theta \qquad (6\text{-}21)$$

The triangular waveform shown in Fig. 6-25d is the integral of the rectangular distribution of the armature current, and the net mmf is the sum of the two mmf distributions. The simplest way to describe the relative position of the two fields is in terms of the fundamental component of the two waveforms, which are 90 electrical degrees apart. The resultant distribution shown in Fig. 6-25e shows that there must be an increase in the flux density at one side of each pole face and a decrease at the other. The nonlinear magnetization characteristic of the pole material has the effect that the increase is less than the decrease. The net result is that there is a decrease in the flux per pole as the armature current increases. While this reduction in flux is normally only a few percent, the effect can be quite pronounced, as illustrated in Fig. 6-22. Measurement of the armature reaction effect is not easy and is imprecise due to magnetic hysteresis. The overall effect is that the terminal voltage predicted for a generator will be higher than it should be, and the speed predicted for a motor will be lower than it should be when this effect is neglected.

The main effect of armature reaction can now be seen to be similar to that of a differential compound winding. Since the characteristics of normal motors and generators are adversely affected by armature reaction, it should now be evident that differential compounding can be expected to produce similar results.

The armature reaction also shifts the points in the air gap where the magnetic field is zero. This means that unless the brushes are shifted to the new position of zero field, they will short-circuit at least one armature coil while there is a voltage induced in it. Rather than move the brushes, it is preferable to distort the field further by means of *interpoles* or *commutating poles*. These are placed between the main poles and carry the armature current. In addition to cancelling the field due to the armature reaction, they are usually adjusted to produce sufficient flux density in this part of the gap to induce a voltage equal and opposite to the self-induced voltage resulting from the rapid change in the coil current during commutation. Other than including the resistance of the interpole winding with the armature

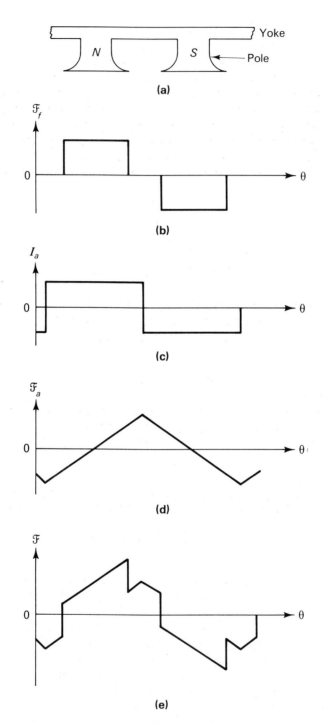

Figure 6-25 Armature and field MMF Distribution.

resistance, there is no need to alter the model, since the interpoles have the effect of making the real dc machine behave more closely to the ideal.

6.7 CONTROL OF DC MOTORS

From the examples in the previous sections it should now be evident that it is a relatively simple matter to control the speed of a dc motor. It is possible to draw some general conclusions by rearranging some of the equations. From Eqs. (6-14) and (6-16) the speed can be expressed as

$$\omega_m = \frac{V_t - R_a I_a}{K_v \phi} \tag{6-22}$$

If we assume that the motor is unsaturated, the flux can be taken as directly proportional to the field current so that

$$\phi = K_f I_f \tag{6-23}$$

and the speed is given directly as

$$\omega_m = \frac{V_t - R_a I_a}{K_v K_f I_f} \tag{6-24}$$

The field current may be supplied from a separate source having a terminal voltage V_f, so that Eq. (6-24) becomes

$$\omega_m = \frac{V_t - R_a I_a}{K_v K_f \dfrac{V_f}{R_f}} \tag{6-25}$$

It is evident from this expression that there are several methods by which the speed can be varied.

1. The voltage applied to either the armature or field circuit can be varied. That is, the speed is controlled by changing V_t or V_f.
2. Additional resistance can be inserted in the armature circuit as illustrated in Example 6-3. However, this method is inefficient and not very effective at light loads. Another disadvantage is that often the series resistance may have a limited number of segments, resulting in discontinuous control.
3. Additional resistance can be inserted in the field circuit as illustrated in Example 6-4. This is relatively efficient, but not as convenient nor as efficient as controlling the terminal voltage directly.

With the development of power semiconductor devices and power electronics technology, a variable dc voltage can be obtained from a fixed dc voltage source by means of a dc chopper. When an ac source is available, the variable dc voltage is obtained by means of a *converter* or controlled rectifier. The efficiency of such power electronics circuits is very high, normally at least 98%, so that the overall efficiency of such a drive system is high. In practice, two controllers are often used: One controls the armature voltage and is used for speeds below base speed; the other controls the field voltage and is used for speeds above base speed. It is also possible to reverse the power flow through the power controller so that regenerative braking can be obtained.

Perhaps the most common semiconductor device used for power control is the thyristor. This is a three-terminal device (the terminals are called anode, cathode, and gate) which has the property that the instant during a cycle at which it begins to conduct is controlled by applying a pulse to the gate. In its simplest form, the current cannot be switched off by another pulse, but after a point in the waveform at which the current is zero due to the fact that the voltage applied to the circuit is sinusoidal, the current remains zero and the thyristor is effectively turned off. Since this is the natural behavior of an ac circuit, the process is called *natural commutation*. Naturally commutated thyristors are normally used in ac–dc converters.

The semiconductor devices used in choppers may be power transistors, forced-commutated thyristors, or gate turn-off thyristors (GTOs). A power transistor has three terminals (collector, emitter, and base) and is normally operated as a switch by applying a control voltage between the base and the emitter. A forced-commutated thyristor is switched on by a pulse and is turned off by an auxiliary commutation circuit. A GTO is turned on by means of a positive pulse and is turned off by a negative pulse. Typical switching times for turning on these devices are of the order of 1 μs. The corresponding times to turn off thyristors are of the order of 10 μs, but for transistors these are likely to be much less.

6.7.1 DC Chopper Control

In essence, a *chopper* acts as a switch that is turned on and off in regular cycles. The operation can be explained by referring to Fig. 6-26, which shows a chopper connected to an ideal resistive load. If t_1 is the time in one cycle for which it is conducting and t_2 is the time for which it blocks conduction, the period is

$$T = t_1 + t_2 \tag{6-26}$$

The average voltage across the load is simply the average value of this waveform of amplitude V_s, the source voltage. Thus the terminal voltage at the resistive load is

$$V_t = \frac{t_1}{T}V_s = \frac{t_1}{t_1 + t_2}V_s = \delta V_s \tag{6-27}$$

where δ is the duty cycle. Thus, by varying δ, the terminal voltage can be varied between zero and V_s.

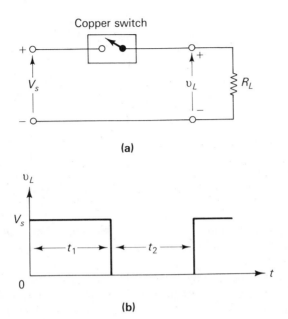

Figure 6-26   DC Chopper.

In Example 6-3, the speed voltage required for operation at 600 r/min was found to be 420 V with an armature current of 171.4 A. With chopper control, the voltage at the terminals of the armature is therefore

$$V_t = 420 + 0.05 \times 171.4 = 428.57 \text{ V}$$

Since the source voltage is 500 V, the duty cycle of the chopper is

$$\delta = \frac{428.57}{500} = 0.857 \text{ or } 85.7\%$$

In Example 6-4, the field current required for operation at 2000 r/min was found to be 1.88 A, so that the voltage required across the field winding is

$$V_f = 200 \times 1.88 = 376 \text{ V}$$

With the source voltage of 500 V, the duty cycle required is

$$\delta = \frac{376}{500} = 0.752 \text{ or } 75.2\%$$

6.7.2 Single-Phase Full-Wave Converter

This form of control requires a sinusoidal source, and naturally commutated thyristors are normally used. A schematic diagram for an ideal resistive load is shown in Fig. 6-27, and the voltage waveforms are shown in Fig. 6-28. The pulses applied to the gates are arranged to have T_1 and

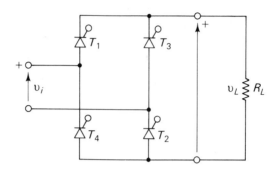

Figure 6-27   Single Phase Full Wave Converter.

T_2 conduct while T_3 and T_4 are in their blocking state. One-half period later it is thyristors T_3 and T_4 that conduct while T_1 and T_2 block the flow of current.

The delay in permitting conduction is usually expressed as the *delay angle* or *firing angle,* α. When α is zero, the converter behaves as if it were an ordinary full-wave rectifier, but by increasing it to 180° the output voltage may be decreased to zero. The average value of the output voltage is given by

$$V_t = \frac{1}{\pi} \int_\alpha^\pi \sqrt{2} V_s \sin \theta \, d\theta = \frac{\sqrt{2} V_s}{\pi} (1 + \cos \alpha) \qquad (6\text{-}28)$$

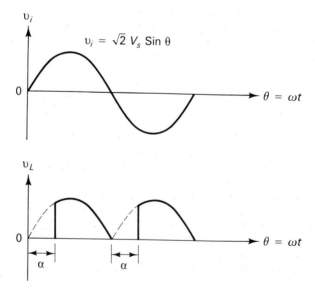

Figure 6-28   Voltage Waveforms.

In Example 6-3, if the 428.57-V dc is to be obtained from a 550-V, 60-Hz single-phase source, then the delay angle is obtained from

$$428.57 = \frac{\sqrt{2} \times 550}{\pi}(1 + \cos \alpha)$$

$$\cos \alpha = \pi \times \frac{428.57}{\sqrt{2} \times 550} - 1 = 0.73$$

$$\alpha = 43°$$

In Example 6-4, if the 376-V dc is to be obtained from a similar source, the delay angle is obtained from

$$376 = \frac{\sqrt{2} \times 550}{\pi}(1 + \cos \alpha)$$

$$\cos \alpha = \frac{376\pi}{\sqrt{2} \times 550} - 1 = 0.5186$$

$$\alpha = 58.8°$$

Note that for higher power applications a three-phase full-wave converter is normally used. Also, in each of these calculations the inductance has been neglected, as has the speed voltage with the armature voltage control. These modify the waveforms so that the results obtained above are approximate. However, the principles are valid, and once the correct waveforms have been obtained the procedure is the same.

PROBLEMS

6-1. The armature of a separately excited dc machine is connected to a 220-V source, and it rotates at 1110 r/min. The cross-sectional area of the face of each of its four poles is 200 cm^2, and the average flux density in the air gap under each pole face is 0.75 T. The machine constants are:

$$K_a = 115 \text{ V·s/Wb}$$

$$K_t = 115 \text{ N·m/(Wb·A)}$$

The resistance of the armature winding, R_a, is 0.12 Ω. All losses and the effect of armature reaction are to be neglected.
(a) Is the machine operating as a motor or as a generator?
(b) Calculate the developed power.
(c) Calculate the developed torque.

6-2. A 750-kW, dc separately excited generator has a terminal voltage of 600 V at rated current. It is being driven at a speed of 500 r/min. The armature resistance is 0.008 Ω. The rated field current is 12 A and the field resistance is 25 Ω. The mechanical loss at 500 r/min is 10 kW. The iron loss at the rated excitation and at 500 r/min is 12 kW. The effects of armature reaction are to be neglected, and the stray and brush losses are also negligible.
(a) Calculate the speed voltage when delivering rated armature current at 600 V.

(b) Calculate the mechanical power requirement to drive the generator.
(c) Calculate the shaft torque.
(d) Calculate the voltage regulation at rated current.
(e) Calculate the efficiency for this load.

6-3. A self-excited dc shunt generator has the following magnetization characteristic at 900 r/min.

I_f	0	0.92	1.26	1.54	1.65	1.85	2.05	2.44	2.97	3.60
E_a	8	160	210	240	250	265	280	300	320	340

The armature resistance is 0.3 Ω and the field resistance is 125 Ω.
(a) Determine the resistance to be connected in series with the field winding to result in an open circuit voltage of 250 V when it is driven at 900 r/min.
(b) With this resistance in the field circuit, but neglecting the effect of armature reaction, determine the terminal voltage when the armature current is 50 A.
(c) Determine whether or not it is possible to reduce the resistance in series with the field winding so that the terminal voltage with 50-A armature current is 250 V. If so, calculate this value of resistance.

6-4. A certain dc separately excited motor has an armature resistance of 0.012 Ω and a field resistance of 15 Ω. Its field winding is connected to a 450-V source. When its armature is connected to a 600-V source, its no-load speed is 430 r/min with an armature current of 84 A. A load is applied to the shaft such that the speed drops to 410 r/min. Neglecting the effect of armature reaction, determine the corresponding efficiency.

6-5. A dc separately excited motor has an armature resistance of 0.032 Ω and a field resistance of 25 Ω. Its field winding is connected to a 400-V source. When its armature is connected to a 500-V source, its no-load speed is 600 r/min with an armature current of 50 A. A load is applied to the shaft such that the speed drops to 580 r/min. Neglecting the effect of armature reaction, determine the corresponding efficiency.

6-6. A separately excited dc motor forms part of an adjustable-speed-drive system. Its field current is constant and its armature is connected to a variable voltage source. When this voltage is set to 500 V, the no-load speed is 1050 r/min. For this problem, all losses and armature reaction are to be ignored.
(a) Calculate the armature current required to develop a torque of 300 N·m.
(b) The speed is to be increased to 2000 r/min, with the armature voltage limited to 500 V and the armature current limited to that found in part (a). Determine the maximum value of developed torque that may be obtained at this speed, assuming that the flux is proportional to the field current.

6-7. The magnetization characteristic of a certain dc machine at 1200 r/min is as follows.

I_f	0	1.0	2.0	3.0	4.0	5.0	6.0	7.0	8.0	9.0
E_a	5	50.4	92.5	128.3	159.1	185.8	209.3	230.1	248.6	265.1

The armature resistance voltage drop at no-load is negligible, and the field winding resistance is 33.0 Ω. The machine is to be connected and used as a shunt motor.

(a) Determine the no-load speed when connected to a 220-V source, there being no additional resistance connected in series with the field winding.

(b) Determine the value of resistance which should be connected in series with the field winding to give a no-load speed of 1450 r/min.

(c) What percentage change in speed will result when the terminal voltage is reduced by 10%, with the field resistance restored to that in part (a)?

(d) What percentage change in speed will result when the terminal voltage is increased by 10%, with the field resistance restored to that in part (a)?

6-8. The magnetization characteristic of a dc shunt motor at 1150 r/min is as follows.

I_f	0	1.0	2.0	3.0	4.0	5.0	6.0	7.0	8.0	9.0
E_a	9	93.2	171.1	237.4	294.3	343.7	387.2	425.7	459.9	490.4

The armature resistance voltage drop at no-load is negligible and the field winding resistance is 61.0 Ω. The machine is to be connected and used as a shunt motor.

(a) Determine the no-load speed when both armature and field are connected directly to a 400-V source.

(b) Determine the value of resistance that should be connected in series with the field winding to give a no-load speed of 1200 r/min.

(c) What percentage change in speed will result when the terminal voltage is reduced by 10%, with the field resistance restored to that in part (a)?

(d) What percentage change in speed will result when the terminal voltage is increased by 10%, with the field resistance restored to that in part (a)?

6-9. A certain 240-V dc shunt motor has an armature circuit resistance of 0.11 Ω, and the shunt-field resistance is 120 Ω. When connected to a 240-V source and driving a particular load, the line current is 82 A and the speed is 900 r/min. Neglecting the effect of armature reaction, determine the armature current and speed if the load torque is increased by 50%.

Calculate the percentage change in flux per pole required to restore the speed to the original value of 900 r/min at this increased load torque.

6-10. The armature circuit resistance of a 500-V dc shunt motor is 0.2 Ω, and its shunt-field resistance is 250 Ω. When connected to a 500-V source and driving a particular load, the total current is 100 A and the speed is 1100 r/min. Neglecting the effect of armature reaction, determine the armature current and speed if the load torque is doubled.

Calculate the percentage change in flux per pole required to restore the speed to the original value of 1100 r/min at this increased load torque.

6-11. A dc shunt motor is driving a mechanical system that requires a torque proportional to speed. Initially it is running at 1000 r/min with a current of 100 A when connected to a 500-V supply. The resistance of the armature is 0.2 Ω and that of the field winding is 250 Ω.

(a) The speed is to be reduced to 800 r/min by means of a resistance inserted in series with the armature, the field current being kept constant. Determine the value of resistance required.

(b) If the speed is to be increased to 1200 r/min, with the armature again connected directly to the source, find the resistance that should be connected in series with the field winding. The flux per pole may be assumed proportional to the field current.

6-12. A 550-V dc shunt motor has an armature circuit resistance of 0.15 Ω and a field resistance of 261 Ω. When driving a mechanical system whose torque requirements are independent of speed, the armature current is 30 A and the speed is 1500 r/min when it is connected to a 550-V source. The effect of armature reaction is to be neglected.

(a) Determine the no-load speed.

(b) Determine the speed at which the motor would drive the system when a resistance of 0.55 Ω is inserted in series with the armature, the terminal voltage remaining at 550 V.

6-13. A dc shunt motor drives a mechanical system requiring torque proportional to speed. When connected to a 500-V source, it runs at a speed of 1200 r/min with an armature current of 120 A and a field current of 6.5 A. The armature resistance is 0.15 Ω and the effect of armature reaction is to be neglected.

(a) Determine the resistance that would reduce the speed to 1000 r/min when connected in series with the armature.

(b) Determine the resistance that would increase the speed to 1500 r/min when connected in series with the field winding. Assume flux is proportional to field current.

6-14. A dc shunt motor drives a mechanical system requiring torque independent of speed. When connected to a 750-V source, it runs at a speed of 1400 r/min with an armature current of 200 A and a field current of 8.3 A. The armature resistance is 0.05 Ω and the effect of armature reaction is to be neglected. Assuming that suitable converters are available, determine

(a) the average voltage to be applied to the armature to give a speed of 1000 r/min,

(b) the average voltage to be connected to the field winding to give a speed of 1800 r/min. Assume flux is proportional to field current.

6-15. The armature resistance of a dc shunt motor is 0.25 Ω. When driving a certain system, the armature current is 50 A and the speed is 850 r/min when the armature terminal voltage is 220 V. Mechanical losses may be neglected.

(a) Calculate the shaft torque.

(b) Calculate the speed if the flux is reduced by 10% and the load torque remains constant.

6-16. A dc motor has the following open circuit characteristic at 1150 r/min.

I_f	0.5	1.0	1.5	2.0	2.5	3.0	3.5	4.0
E_a	92	184	268	363	442	478	505	520

If the field current is 3.0 A and the armature current is 100 A, calculate the value of the developed torque.

6-17. A dc series motor operates at 750 r/min with a line current of 100 A from a 230-V source. Its armature resistance is 0.1 Ω and its field resistance is 0.07 Ω.

Assuming that the flux corresponding to a current of 50 A is 90% of that corresponding to a current of 100 A, find the motor speed at a line current of 50 A when the terminal voltage is 230 V.

6-18. A series motor is to drive a system that requires a constant torque of 25 N·m. The resistance of its armature and field circuit is 0.4 Ω, and a 220-V dc source is available. With its shaft held stationary, the motor was found to produce a torque of 15 N·m with a current of 20 A.

Assuming that it will operate with its magnetic field system unsaturated, determine the speed at which it will drive the load.

6-19. A dc series motor has the following magnetization characteristic at 700 r/min.

I_f	50	100	150	200	250	300	350	400
E_a	129	258	376	510	620	671	709	730

The resistance of its armature is 0.01 Ω and that of its field is 0.004 Ω. When connected to a 600-V source, it drives a particular mechanical system at 650 r/min. Find the current and the developed torque. (Hint: Plot the speed/current characteristic or use an iterative approach.)

7

INDUCTION MACHINES

7.1 INTRODUCTION

The most common alternating current machine in general industrial use is the three-phase induction motor. It is available in sizes of less than 1 kW to more than 1000 kW. In one type there are no sliding contacts such as at the commutator of a dc machine. The other type requires slip rings, which require less maintenance than the commutator of a dc machine. Figure 7-1 shows a general view of a wound rotor having a set of three slip rings. A close-up view of a set of slip rings with the brushes in position is shown in Fig. 7-2, and some details of the brush-holder assembly can be seen in Fig. 7-3. The presence of the slip rings can often be noticed because of the housing required at one end of the motor, as illustrated in Fig. 7-4.

The original invention of the induction motor is usually attributed to Nicholas Tesla in the late nineteenth century; work is still being done to improve the basic machine and to incorporate it in variable-speed-drive systems. The treatment that follows is based on a simple model of the structure and the phenomena taking place in it. For the purpose of applications of induction motors, the standard circuit model is usually sufficient, although there are more detailed models that are sometimes required.

The excitation of an induction motor is provided by a polyphase winding, which is normally connected to a balanced polyphase source. Normally a three-phase system is used, but two-phase induction motors are to be found, usually in older automatic-control systems. With the

Figure 7-1   3-Phase Wound Rotor (Photo courtesy of Canadian General Electric Co.).

Figure 7-2   Brush Holders Assembled to Collector Rings (Photo courtesy of Canadian General Electric Co.).

Figure 7-3   Brush Holder Assembly (Photo courtesy of Canadian General Electric Co.).

Figure 7-4   Wound Rotor Induction Motor (Photo courtesy of Canadian General Electric Co.).

development of thyristor circuits it is now common to find three-phase motors in drive systems that form part of a control system. A recent development is the six-phase motor, which is always fed by means of a power electronic circuit. It is, strictly speaking, half of a 12-phase winding, just as an ordinary two-phase system is half of a four-phase system and is being considered for applications where reliability is particularly important.

For applications where the power required is small, or where only a single-phase source is available, a single-phase version of the induction motor can be used. It is to be found in many domestic appliances, such as washers, dryers, air-conditioning units, and fans.

7.2 THE REVOLVING FIELD

At the heart of the operation of all ac machines is the concept of the rotating field produced by a set of static coils. To explain this concept, the algebra is reduced considerably if we consider a two-phase system. The main point is that any balanced q-phase winding carrying balanced q-phase currents will produce a rotating field.

Consider the two-phase winding shown schematically in Fig. 7-5. The complete winding consists of two sets of coils mounted on the stator and forming the phase windings. Each phase is shown as only one coil, arranged such that its axis of magnetization is displaced by 90° from that of the other phase. Each coil produces a field pattern which is sinusoidally distributed in the space of the air gap between the rotor and the stator. That is, the mmf distribution of coil x is

$$\mathscr{F}_x(\theta) = N_x i_x \cos \theta \tag{7-1}$$

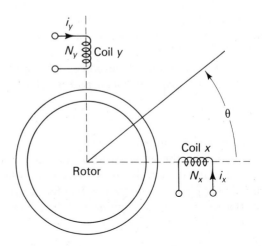

Figure 7-5 Schematic Diagram of 2-Phase Winding.

where N_x is the number of turns in coil x, i_x is the current in coil x (I_x cos ωt), and θ is the angular position of any point in the air gap.

This may seem a rather idealistic property of a set of coils, and it is true that it can never be fully attained in practice. However, a normal winding comes sufficiently close to this ideal that there is surprisingly little error in assuming this sinusoidal field distribution. The other set of coils produces a similar field distribution about its axis, which is fixed 90° from that of winding x. Since we can have only one variable for position in the air gap, θ, the mmf distribution of this coil is given by

$$\mathscr{F}_y(\theta) = N_y i_y \sin \theta \qquad (7\text{-}2)$$

where N_y is the number of turns in winding y and i_y is the current in winding y ($I_y \sin \omega t$) since the assumption is that the time relationship between the two currents must be the same as the space relationship between the two windings.

In a balanced winding, each coil has the same number of turns ($N_s = N_x = N_y$) and the magnitude of the two currents is equal when the currents are balanced ($I_s = I_x = I_y$) so that the total mmf produced by the stator winding is

$$\mathscr{F}_s(\theta, t) = \mathscr{F}_x + \mathscr{F}_y$$
$$= N_s I_s (\cos \theta \cos \omega t + \sin \theta \sin \omega t)$$
$$= N_s I_s \cos (\theta - \omega t) \qquad (7\text{-}3)$$

This may be recognized as an expression of the function cos θ traveling in the positive direction with (angular) velocity ω. If the radial length of the air gap is uniform—that is, both the stator and rotor have cylindrical surfaces and are concentric—the flux density of the field is also sinusoidal and rotating with angular velocity ω. In Sec. 6-2 it was observed that it is possible to repeat the pole structure several times, thus producing several periods of field distribution. It is also possible to arrange the distribution of this winding so that it produces several cycles of mmf around the air gap, thus producing the effect of more poles. The angles in Eqs. (7-1) to (7-3) are all "electrical," so that the actual velocity of the rotating field is reduced by having multiple poles. The speed at which the field rotates is called the *synchronous speed* and is related to the supply frequency by

$$\omega_s = \frac{2\omega}{P} \qquad (7\text{-}4)$$

where P is the number of poles.

Although this is shown for angular velocity in radians per second, it may be applied directly for speeds in revolutions per minute. Thus with a 60-Hz supply, a two-pole winding has a synchronous speed of 3600 r/min and that of a six-pole winding is 1200 r/min.

7.3 EQUIVALENT CIRCUIT OF THE INDUCTION MACHINE

In keeping with the development of the rotating field of a two-phase winding in the previous section, the circuit model of a balanced induction machine will now be developed for the two-phase case. The conclusions are, of course, applicable to a balanced machine having any number of phases. Thus our machine will have two windings on the stator, (x and y), and two windings on the rotor (a and b), as shown in Fig. 7-6. The rotor windings are short-circuited in normal operation, thus producing maximum induced current.

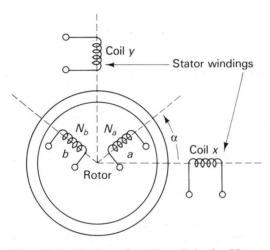

Figure 7-6 Windings of a 2-Phase Induction Motor.

The mmf produced by the stator winding is that given by Eq. (7-3), and this will produce a sinusoidal distribution of flux density since the induction machine has a uniform air gap. As long as the rotor is not rotating at synchronous speed, there will be relative motion between the coils of the rotor winding and the rotating field, with a resulting induced voltage and current. The flux linking winding a at position α is

$$\phi(\alpha, t) = \phi_m \cos(\alpha - \omega t) \qquad (7\text{-}5)$$

so that the voltage induced is

$$e_a(t) = \frac{d}{dt}[N_a \phi_m \cos(\alpha_a - \omega t)]$$

where α_a equals $\omega_m t + \delta$, ω_m is the angular speed of the rotor, and δ is the relative position of the rotor.

For coil b the induced voltage is obtained by noting that the angle α_a above is replaced by

$$\alpha_b = \omega_m t + \delta + 90°$$

Substituting these values in Eq. (7-5), the expressions for the two rotor voltages are

$$e_a(t) = N_a \phi_m \frac{d}{dt}[\cos(\omega_m t + \delta - \omega t)]$$

$$= -N_a \phi_m (\omega - \omega_m) \sin\{(\omega - \omega_m)t - \delta\} \tag{7-6}$$

$$= -sE_m \sin(s\omega t - \delta)$$

and

$$e_b(t) = sE_m \cos(s\omega\, t - \delta) \tag{7-7}$$

since $N_r = N_a = N_b$ is the number of turns on each rotor phase, s is the *slip*, defined by

$$s = \frac{\omega_s - \omega_m}{\omega_s} \tag{7-8}$$

and E_m is the peak value of the voltage induced in each coil when the rotor speed is zero. Note that both the magnitude and frequency of the rotor voltages are dependent on speed. They form a two-phase set which produces a two-phase set of rotor currents, thus producing an mmf pattern which rotates relative to the rotor at angular speed $s\omega_s$. When the expressions are developed it can be shown that the rotation is such that the angular speed of the rotor mmf relative to the stator is exactly equal to that of the mmf produced by the stator currents. Thus the rotor mmf is stationary relative to the stator mmf, just as in the dc machine.

Although the two mmf patterns are stationary with respect to each other, it is more convenient to seek a circuit model by noting the similarity to a transformer where coils are linked with sinusoidally varying fluxes. In this case, the frequencies in the primary and secondary are not equal and it is necessary to modify the secondary circuit so that it is not frequency dependent. This is done by considering that the voltages e_a and e_b form a two-phase source. As with a balanced polyphase transformer, only one phase of the circuit model need be shown. Figure 7-7 shows the equivalent circuit of the rotor when it is regarded as the secondary winding of a transformer.

The secondary current may be obtained using the normal phasor relationships.

$$\hat{I}_2 = \frac{sE_2}{R_2 + jsX_2} \tag{7-9}$$

where R_2 is the total resistance of the secondary circuit, X_2 is the leakage reactance of the secondary circuit at the frequency of the source, and E_2

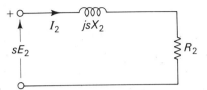

Figure 7-7 Circuit Model of Rotor.

is the rms value of the secondary induced voltage when the speed is zero (at $s = 1$, so that the frequency is that of the supply).

The frequency of this current is that of the source in the rotor, namely slip frequency, and in this form the expression is not particularly useful. It may be changed into one involving a current of the same magnitude but at stator source frequency by dividing numerator and denominator by the slip.

$$\hat{I}_2 = \frac{E_2}{\dfrac{R_2}{s} + jX_2} \tag{7-10}$$

Since the mmf produced by the rotor currents is rotating at synchronous speed relative to the stator winding, it induces a source frequency voltage just as in a normal transformer. It is therefore permissible to connect a circuit whose equilibrium is described by Eq. (7-10) to the secondary terminals of an ideal transformer where the frequency is that of the stator throughout it. This is shown in Fig. 7-8.

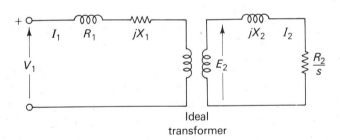

Ideal
transformer

Figure 7-8 Circuit Model of Induction Motor.

As is done when modeling an ordinary transformer, the secondary parameters and variables of this model can be referred to the primary. For convenience, this is now done by adding primes to the symbols for the secondary resistance, leakage reactance, current, and voltage. In addition, shunting branches are added to include the finite permeance and losses of the core. In the case of the induction motor, these losses are those of only the stator, since there is no rotor core loss at synchronous speed. The model correctly predicts conditions at synchronous speed since R_2'/s becomes

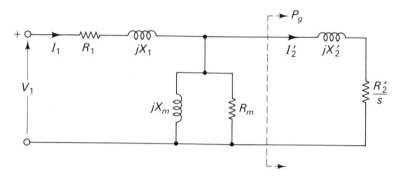

Figure 7-9   Circuit Model Referred to Stator.

infinite and the secondary current must then be zero. The complete circuit model, referred to the primary, is shown in Fig. 7-9.

At this point we must consider the significance of the resistance R_2'/s, which clearly is derived from the actual resistance of the secondary circuit but is significantly different at all values of speed other than zero. Examination of the power flow in the motor provides the answer. The current I_1 evidently is the actual primary current, and the power dissipated in R_1 is the primary copper loss, P_1. If the power dissipated in R_m is the core loss, then the power that remains must be that crossing the air gap and entering the rotor system. This power is usually called the *gap power* and in a *q*-phase motor is given by

$$P_g = qI_2'^2 \frac{R_2'}{s} \text{ W} \tag{7-11}$$

We have already noted that the current I_2' correctly gives the magnitude of the secondary current, and therefore the secondary copper loss is

$$P_2 = qI_2'^2 R_2' \text{ W} \tag{7-12}$$

Since this accounts for all the known losses, the remaining power must be the gross developed power produced by the motor, given by

$$P_d = P_g - P_2$$

$$= qI_2'^2 \left(\frac{R_2'}{s} - R_2' \right)$$

$$= qI_2'^2 \frac{R_2'}{s} (1 - s)$$

$$= P_g(1 - s) \tag{7-13}$$

This may now be equated to the product of developed torque and angular speed to obtain an expression for the developed torque.

$$T_d \omega_m = P_g(1 - s)$$

The angular speed in this expression may be replaced by recalling that the slip is defined by Eq. (7-8) so that

$$T_d \omega_s (1 - s) = P_g (1 - s)$$

and

$$T_d = \frac{P_g}{\omega_s} \tag{7-14}$$

Perhaps this should not be too surprising since it is basically the field rotating at synchronous speed which produces the mechanical power. It is simply the same point that was observed in the dc machine where Eq. (6-11) demonstrated that the corresponding power $(E_a I_a)$ equals the gross mechanical power. Certainly it means that the calculation of developed torque (in newton-meters) is simply a matter of calculating the gap power as if this were a static circuit problem involving a power transformer and then dividing it by the synchronous speed in (mechanical) radians per second.

Before proceeding with the analysis of the circuit model, we must consider the mechanical losses. In the dc machine these consist of the core, windage, and friction losses, and they exist only when the machine is rotating. In the induction machine the stator core loss exists as long as the machine is connected to a source, the rotor core loss exists when the slip is any value other than zero, and the windage and friction losses exist when the machine rotates. Since the windage and friction losses are significantly greater than the others, it is common to group all of them as rotational losses and remove R_m from the circuit model. This is convenient since the no-load power is the sum of these losses other than the small copper losses, and normally it is not possible to separate the core loss from the mechanical loss. Naturally this introduces a small error in computed values of gap power, but this error is quite insignificant. When this is done, the equivalent circuit now appears as shown in Fig. 7-10. When there is no external load applied to the shaft of the motor, the slip adjusts to a small value such that the torque developed is sufficient to supply the no-

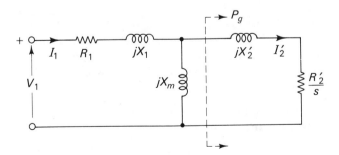

Figure 7-10   Circuit Model, Core Losses among Rotational Losses.

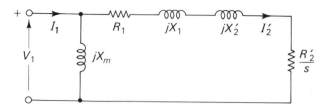

Figure 7-11 Approximate Circuit Model.

load mechanical losses. Thus, in a normal motor, R_2'/s on no-load has a finite value.

Since the circuit model of an induction motor is the same as that of a short-circuited transformer in which the resistance of the secondary circuit is divided by the slip, there may exist the same simplification of the model to reduce the effort within acceptable limits of error. Since this is a question of judgment, there are varying opinions, and therefore we must consider all of the standard approaches. The main problem is that the magnetizing reactance, X_m, is relatively not as large as that of a transformer. The error in displacing the magnetizing reactance to the input terminals as shown in Fig. 7-11 is therefore greater than that for the approximate circuit model of a transformer. At some values of speed the error may be acceptable, but at others this may not be the case.

7.4 ANALYSIS OF EQUIVALENT CIRCUIT

There are several ways by which this circuit may be analyzed, and the choice of method is likely to be made on the basis of what information is to be determined. Two methods are commonly used and will be considered in detail. These are

1. Equivalent impedance of rotor and magnetizing branches,
2. Application of Thevenin's theorem.

The first of these is the more convenient when there is only one operating speed to be considered and the input current is required. The second is valuable when operation over a range of speed is required. In addition, a simplified solution based on the approximate equivalent circuit of Fig. 7-11 may sometimes be acceptable. Other than the different values for the equivalent voltage and impedance, this simplified solution is the same as that using Thevenin's theorem.

7.4.1 Equivalent Impedance of Rotor and Magnetizing Branches

Figure 7-12 shows the circuit model in which the impedance of the two branches has been designated Z_f. Note that this approach is valid only when the stator core loss is included with the rotational losses. The shunting

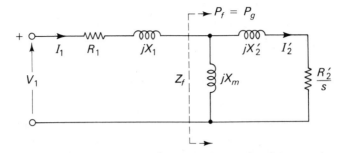

Figure 7-12   Equivalent Circuit of Induction Motor.

branch is therefore purely inductive and the (average) power dissipated in the parallel combination of X_m and $(R_2'/s + jX_2')$ must be the same as the gap power. The fact that one of the parallel impedances is purely imaginary simplifies the calculation of this equivalent impedance. That is,

$$\hat{Z}_f = \frac{\left(\dfrac{R_2'}{s} + jX_2'\right)(jX_m)}{\dfrac{R_2'}{s} + j(X_2' + X_m)}$$

$$= R_f + jX_f$$

(7-15)

where

$$R_f = \frac{X_m^2 \dfrac{R_2'}{s}}{\left(\dfrac{R_2'}{s}\right)^2 + (X_m + X_2')^2}$$

and

$$X_f = \frac{X_m\left[\left(\dfrac{R_2'}{s}\right)^2 + X_2'(X_2' + X_m)\right]}{\left(\dfrac{R_2'}{s}\right)^2 + (X_m + X_2')^2}$$

The input current is now given by

$$\hat{I}_1 = \frac{\hat{V}_1}{(R_1 + R_f) + j(X_1 + X_f)} = I_1 \underline{/\beta}$$

(7-16)

The power dissipated in R_2'/s includes the rotor copper loss, the stator and rotor core losses, and the mechanical losses. Since the average power dissipated in X_m is zero, there is no need to calculate I_2', the gross developed power is given directly by

$$P_g = qI_1^2 R_f$$

and the developed torque is

$$T_d = \frac{P_g}{\omega_s}$$

The net mechanical output or shaft power is obtained by subtracting the rotor copper loss and the rotational losses from the gap power.

$$P_{sh} = P_g(1 - s) - P_{n-l} \tag{7-17}$$

where P_{n-l} is the power on no-load and includes the core losses.

   This approach is very convenient when only one value of speed is involved and the input current is one of the quantities that is to be calculated. It is, however, of little value when more than one speed is involved, such as when obtaining the characteristics of the machine.

7.4.2 Application of Thevenin's Theorem

   Since there is only one element in the circuit model that depends on the speed, it is convenient to take a Thevenin model of part of the circuit as shown in Fig. 7-13. This approach is also known as the *adjusted voltage* method. Again the core, windage, and friction losses are grouped together, and thus the shunt branch consists of only X_m. The application of Thevenin's theorem to this circuit is quite straightforward.

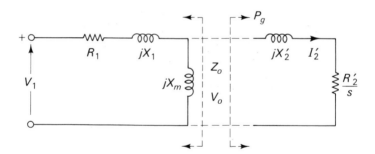

Figure 7-13 Circuit for Thevenin Model.

The variables shown in Fig. 7-13 are given by the following expressions.

$$\hat{V}_o = \frac{jX_m V_1}{R_1 + j(X_1 + X_m)} = V_o \angle \psi \tag{7-18}$$

where

$$V_o = \frac{X_m V_1}{\sqrt{R_1^2 + (X_1 + X_m)^2}}$$

$$\hat{Z}_o = \frac{(R_1 + jX_1)(jX_m)}{R_1 + j(X_1 + X_m)} = R_o + jX_o \tag{7-19}$$

where

$$R_o = \frac{R_1 X_m^2}{R_1^2 + (X_1 + X_m)^2}$$

and

$$X_o = \frac{X_m[R_1^2 + X_1(X_1 + X_m)]}{R_1^2 + (X_1 + X_m)^2}$$

$$\hat{I}_2' = \frac{\hat{V}_o}{\dfrac{R_2'}{s} + R_o + j(X_2' + X_o)} = I_2' \,\underline{/\beta} \qquad (7\text{-}20)$$

where

$$I_2' = \frac{V_o}{\sqrt{\left(\dfrac{R_2'}{s} + R_o\right)^2 + (X_2' + X_o)^2}}$$

To get the gap power, we require the square of the magnitude of the current,

$$I_2'^2 = \frac{V_o^2}{\left(\dfrac{R_2'}{s} + R_o\right)^2 + (X_2' + X_o)^2} \qquad (7\text{-}21)$$

$$P_g = qI_2'^2 \frac{R_2'}{s}$$

$$= \left[\frac{qV_o^2}{\left(\dfrac{R_2'}{s} + R_o\right)^2 + (X_2' + X_o)^2}\right]\frac{R_2'}{s} \qquad (7\text{-}22)$$

$$T_d = \frac{P_g}{\omega_s} \qquad (7\text{-}23)$$

The developed torque can now be plotted as a function of slip or speed as shown in Fig. 7-14, where both speed and slip can be indicated simultaneously.

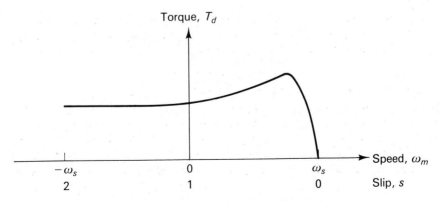

Figure 7-14 Torque-Speed Characteristic.

The developed torque has a clearly defined maximum, generally called the *pull-out torque*. The slip at which this takes place can readily be found by means of the maximum power transfer theorem, since developed torque is obtained by dividing the gap power by the synchronous angular speed. In this case the variable resistance is R_2'/s, so the maximum power occurs when

$$\frac{R_2'}{s} = \sqrt{R_o^2 + (X_2' + X_o)^2} \tag{7-24}$$

and substituting this expression for R_2'/s at maximum developed torque in Eq. (7-22) gives the maximum gap power as

$$P_{gm} = \frac{qV_o^2 \sqrt{R_o^2 + (X_2' + X_o)^2}}{[\sqrt{R_o^2 + (X_2' + X_o)^2} + R_o]^2 + (X_2' + X_o)^2}$$

This can be somewhat simplified, and the expression for maximum developed power is

$$P_{gm} = \frac{qV_o^2}{2[R_o + \sqrt{R_o^2 + (X_2' + X_o)^2}]} \tag{7-25}$$

Dividing by the synchronous speed gives the maximum developed torque as

$$T_{dm} = \frac{qV_o^2}{2\omega_s[R_o + \sqrt{R_o^2 + (X_2' + X_o)^2}]} \tag{7-26}$$

Clearly these expressions give little indication of the form of the characteristics. To do this, we may now introduce approximations based on the relative values of stator resistance and the impedance of the rest of the circuit, which is normally much larger. When the stator resistance is neglected, R_o becomes zero and the corresponding expressions become

$$\frac{R_2'}{s} = X_2' + X_o \text{ (at maximum torque)}$$

$$T_{dm} = \frac{qV_o^2}{2\omega_s(X_2' + X_o)} \tag{7-27}$$

$$T_d = \frac{qV_o^2}{\omega_s\left[\left(\dfrac{R_2'}{s}\right)^2 + (X_2' + X_o)^2\right]} \frac{R_2'}{s} \tag{7-28}$$

If we let s_p be the slip at maximum torque, that is

$$s_p = \frac{R_2'}{X_2' + X_o} \tag{7-29}$$

the ratio of the torque developed at slip s to the pull-out torque can be shown to be

$$\frac{T_d}{T_{dm}} = \frac{2(X_2' + X_o)}{\left(\dfrac{R_2'}{s}\right)^2 + (X_2' + X_o)^2} \frac{R_2'}{s}$$

$$= \frac{2s_p s}{s_p^2 + s^2} \tag{7-30}$$

When the motor is lightly loaded, the slip is low and normally much less than that at the pull-out torque. In Eq. (7-30), s^2 is negligible in comparison with s_p^2, and the developed torque is proportional to the slip. The speed-torque characteristic of the induction motor for this condition has the same form as that of the dc motor, since Eq. (7-30) becomes

$$T_d = \frac{2T_{dm}}{s_p} s$$

$$= \frac{2T_{dm}}{s_p}\left(\frac{\omega_s - \omega_m}{\omega_s}\right)$$

and

$$\omega_m = \omega_s - \frac{s_p \omega_s}{2T_{dm}} T_d \tag{7-31}$$

The similarity can be seen by comparing Eq. (7-31) with Eq. (6-17). At this point we may note that the nameplate data for an induction motor will normally include the number of phases, the line voltage, the frequency, the rated power and line current, and the corresponding speed. Since the slip at rated load is normally very small, there is no difficulty in determining the number of poles.

Example 7-1

A three-phase, 2200-V, 60-Hz, six-pole, wye-connected induction motor has the following parameters in ohms per phase.

$$R_1 = 3.5 \qquad X_1 = 5.8$$
$$R_2' = 2.4 \qquad X_2' = 5.8$$
$$X_m = 300$$

The core, windage, and friction losses are 800 W and may be assumed constant. For a motor speed of 1176 r/min, use the equivalent rotor and magnetizing impedance method to calculate

(a) the input current and power factor,

(b) the developed torque,

(c) the shaft power,

(d) the efficiency,

(e) the starting current and developed torque if connected directly to the 2200-V supply.

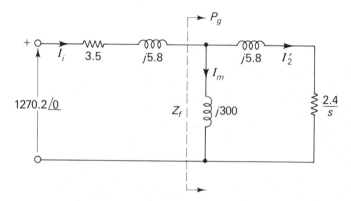

Figure 7-15 Circuit Model, Example 7-1.

Solution. The circuit model is shown in Fig. 7-15 with the line-to-neutral voltage at the source, corresponding to the wye connection. Since the motor has six poles, the synchronous speed is 1200 r/min, and thus the operating slip is

$$s = \frac{1200 - 1176}{1200} = 0.02$$

The effective value of rotor resistance in the circuit model is therefore $2.4/0.02 = 120\ \Omega$ per phase.

(a) The equivalent impedance of the rotor and magnetizing branches is

$$\hat{Z}_f = \frac{(120 + j5.8)(0 + j300)}{120 + j305.8}$$

$$= 100.08 + j44.96\ \Omega$$

The input impedance is

$$\hat{Z}_i = 3.5 + j5.8 + 100.08 + j44.96 = 103.58 + j50.76\ \Omega$$

$$\hat{I}_i = \frac{1270.2}{\hat{Z}_i} = 9.888 - j4.846 = 11.01\ \underline{/-26.1}\ A$$

The input (line) current is therefore 11.01 amperes and the power factor is the cosine of $26.1° = 0.898$ (lagging).

(b) the developed torque is obtained by first calculating the air gap power. This is

$$P_g = 3 \times 11.01^2 \times 100.08 = 36\ 395\ W$$

and the synchronous speed is

$$\omega_s = \frac{4\pi f}{P} = 125.66\ \text{rad/s}$$

The developed torque is

$$T_d = \frac{36\ 395}{125.66} = 289.6\ \text{N·m}$$

(c) The shaft power is obtained by subtracting the known losses from the gap power.

$$P_{sh} = P_g - P_2 - P_{n-l}$$
$$= P_g(1 - s) - P_{n-l}$$
$$= 36\ 395 \times (1 - 0.02) - 800$$
$$= 36\ 667 - 800$$
$$= 34\ 867\ \text{W}$$

(d) The efficiency requires the calculation of the stator copper loss in addition to the values of power obtained in part (c). This loss is given by

$$P_1 = 3 \times 11.01^2 \times 3.5 = 1273\ \text{W}$$

The input power is obtained by adding this loss to the gap power and is 37 668 W. Dividing the net output power of 34 867 W by this value gives the efficiency as 0.926 or 92.6%.

(e) In this case the speed is zero, so that the value of the slip is 1.0 and the calculation is simply a repeat of parts (a) and (b). The results are as follows.

$$\hat{Z}_f = 2.3097 + j5.708\ \Omega$$

$$\hat{Z}_i = 5.8097 + j11.508\ \Omega$$

$$\hat{I}_i = 44.4 - j87.96 = 98.5\ \underline{/-63.2}\ \text{A}$$

$$P_g = 3 \times 98.53^2 \times 2.3097 = 67\ 269\ \text{W}$$

$$T_d = \frac{67\ 269}{125.66} = 535.3\ \text{N·m}$$

Example 7-2

For the same motor as that in Example 7-1, use the Thevenin equivalent model to calculate

(a) the input current and power factor,
(b) the developed torque,
(c) the shaft power,
(d) the efficiency,
(e) the starting current and developed torque if connected directly to the 2200-V supply,
(f) the speed at which maximum torque is developed and this value of torque.

Repeat part (f), neglecting the stator resistance.

Solution

(a) Referring to Fig. 7-15, the Thevenin model of the stator and magnetizing branches is obtained by first calculating the equivalent source voltage and impedance. This is simply a matter of substituting the numerical values in Eqs. (7-18) and (7-19). The voltage is

$$\hat{V}_o = (0 + j300) \times \frac{1270.2}{3.5 + j305.8} = 1245.9 + j14.3$$

$$= 1246\ \underline{/0.66}\ \text{V}$$

The impedance is

$$\hat{Z}_o = \frac{(0 + j300)(3.5 + j5.8)}{3.5 + j305.8} = 3.368 + j5.729$$

$$= 6.646 \; \underline{/59.55} \; \Omega$$

The total impedance is

$$\hat{Z} = 3.368 + j5.729 + 120 + j5.8 = 123.9 \; \underline{/5.34} \; \Omega$$

The rotor current is

$$\hat{I}_2' = \frac{1246 \; \underline{/0.66}}{123.9 \; \underline{/5.34}} = 10.05 \; \underline{/-4.68} = 10.02 - j0.82 \text{ A}$$

To get the stator current, we must first obtain the current through the magnetizing branch, relative to the same reference phasor that is the actual input voltage. The voltage across the magnetizing branch is the same as that across the rotor circuit, namely

$$\hat{V}_m = (10.02 - j0.82) \times (120 + j5.8)$$

$$= 1207.2 - j40.3$$

$$= 1207.8 \; \underline{/-1.91} \text{ V}$$

The magnetizing current is therefore

$$\hat{I}_m = \frac{1207.8 \; \underline{/-1.91}}{300 \; \underline{/90}} = 4.026 \; \underline{/-91.91}$$

$$= -0.134 - j4.024 \text{ A}$$

and the stator current is

$$\hat{I}_i = \hat{I}_2' + \hat{I}_m$$

$$= (10.02 - j0.82) + (-0.134 - j4.024)$$

$$= 9.886 - j4.844$$

$$= 11.01 \; \underline{/-26.1} \text{ A}$$

This is clearly inconvenient when the input current is required. However, when only the developed torque is required, there is no need to keep track of the phase angles, as will be seen in the solution to part (b).

(b) When using the Thevenin model, we have the actual rotor current so we may use the actual rotor circuit resistance divided by slip. That is

$$P_g = 3 \times 10.05^2 \times 120$$

$$= 36\ 361 \text{ W}$$

This slight difference from the value of 36 395 W obtained in Example 7-1, using the equivalent impedance of rotor and magnetizing branches, is due to rounding off the intermediate values. Better agreement may be obtained by maintaining a larger number of significant figures throughout the calculations, even though the accuracy of the parameter values normally does not justify it.

(c) The shaft power is obtained by subtracting the known losses from the gap power.

$$P_{sh} = P_g - P_2 - P_{n-l}$$
$$= P_g(1 - s) - P_{n-l}$$
$$= 36\ 361 \times (1 - 0.02) - 800$$
$$= 35\ 634 - 800$$
$$= 34\ 834\ \text{W}$$

(d) The efficiency requires the calculation of the stator copper loss in addition to the values of power obtained in part (c). This is

$$P_1 = 3 \times 11.01^2 \times 3.5 = 1273\ \text{W}$$

The input power is obtained by adding this loss to the gap power and is 37 634 W. Dividing the net output power of 34 834 W by this value gives the efficiency as 0.926 or 92.6%.

(e) With the Thevenin model, the rotor current and developed torque for the starting condition are more readily obtained than in Example 7-1. Repeating parts (a) and (b) with $s = 1.0$ gives the following values directly. The total impedance is

$$\hat{Z} = 3.368 + j5.729 + 2.4 + j5.8 = 5.768 + j11.529 = 12.891\ \underline{/63.42}\ \Omega$$

The rotor current is

$$\hat{I}_2' = \frac{1246\ \underline{/0.66}}{12.891\ \underline{/63.42}} = 96.66\ \underline{/-62.76} = 44.24 - j85.94\ \text{A}$$

The gap power is obtained as

$$P_g = 3 \times 96.66^2 \times 2.4$$
$$= 67\ 271\ \text{W}$$

Since the synchronous speed is 125.66 rad/s, the developed torque on starting is

$$T_d = \frac{67\ 271}{125.66} = 535.3\ \text{N·m}$$

As in part (a), the stator current is obtained by first calculating the current through the magnetizing branch, relative to the same reference phasor. The voltage across the magnetizing branch is the same as that across the rotor circuit, namely

$$\hat{V}_m = (44.24 - j85.94) \times (2.4 + j5.8)$$
$$= 604.58 - j50.24$$
$$= 606.66\ \underline{/-4.75}\ \text{V}$$

The magnetizing current is therefore

$$\hat{I}_m = \frac{606.66\ \underline{/-4.75}}{300\ \underline{/90}} = 2.022\ \underline{/-94.75}$$
$$= -0.167 - j2.015\ \text{A}$$

and the stator current on starting is

$$\hat{I}_i = \hat{I}_2' + \hat{I}_m$$
$$= (44.24 - j85.94) + (-0.167 - j2.015)$$
$$= 44.07 - j87.96$$
$$= 98.38 \;\underline{/-63.39} \; A$$

(f) The condition for maximum developed torque is

$$\frac{R_2'}{s} = |R_o + jX_o + jX_2'|$$
$$= |3.368 + j5.729 + j5.8| = 12.01$$

The slip at maximum torque is therefore $2.4/12.01 = 0.1998$, which realistically must be taken as 0.2; the corresponding speed is 960 r/min. The value of 12.01 may be substituted directly for R_2'/s and the impedance is thus

$$Z = |3.368 + j5.729 + j5.8 + 12.01| = |15.378 + j11.529| = 19.22 \; \Omega$$

The rotor current at maximum torque is

$$I_2' = \frac{1246}{19.22} = 64.83 \; A$$

so that the maximum gap power is

$$P_{gm} = 3 \times 64.83^2 \times 12.01 = 151\,432 \; W$$

and the maximum developed torque is

$$T_{dm} = \frac{151\,432}{125.66} = 1205 \; N{\cdot}m$$

When the stator resistance is neglected, as in Eqs. (7-27) to (7-29), the slip at maximum torque is calculated as 0.209 and the maximum torque is 1613 N·m, which indicates that this approximation must be used with extreme caution. Strictly, this completes the discussion of the two most common models. However, it is instructive to use the approximate model and, by comparing the results, demonstrate the errors introduced at rated load. Of course it must be noted that at lower loads these errors are inevitably greater.

Example 7-3

For the same motor as that in Example 7-1, use the approximate circuit model to calculate

    (a) the input current and power factor,

    (b) the developed torque,

    (c) the shaft power,

    (d) the efficiency,

    (e) the starting current and developed torque if connected directly to the 2200-V supply,

    (f) the speed at which maximum torque is developed and this value of torque.

Solution

(a) When the approximate model shown in Fig. 7-16 is used, the solution is similar to that using the Thevenin model, but simpler. In this case the total impedance of the stator and rotor windings is

$$\hat{Z} = 3.5 + j5.8 + 120 + j5.8 = 123.5 + j11.6 \ \Omega$$

The rotor current is therefore

$$\hat{I}'_2 = \frac{1270.2}{123.5 + j11.6} = 10.195 - j0.958 = 10.24 \ \underline{/-5.4} \text{ A}$$

and the magnetizing current is

$$\hat{I}_m = \frac{1270.2}{0 + j300} = 4.234 \ \underline{/-90} \text{ A}$$

The approximate stator current is the sum of these two currents:

$$\hat{I}_i = 10.195 - j0.958 - j4.234$$
$$= 10.195 - j5.192$$
$$= 11.44 \ \underline{/-27.0} \text{ A}$$

and the power factor is the cosine of $27.0° = 0.891$ (lagging).

(b) When using the approximate model, we have the (approximate) rotor current, so it is the rotor circuit resistance divided by slip which is used directly.

$$P_g = 3 \times 10.24^2 \times 120 = 37\ 749 \text{ W}$$

(c) The shaft power is obtained by subtracting the known losses from the gap power.

$$P_{sh} = P_g - P_2 - P_{n-l}$$
$$= P_g(1 - s) - P_{n-l}$$
$$= 37\ 749 \times (1 - 0.02) - 800$$
$$= 36\ 994 - 800$$
$$= 36\ 194 \text{ W}$$

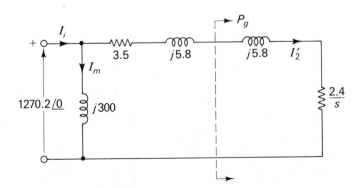

Figure 7-16   Approximate Equivalent Circuit, Example 7-3.

(d) The efficiency requires the calculation of the stator copper loss in addition to the values of power obtained in part (c).

$$P_1 = 3 \times 10.24^2 \times 3.5 = 1101 \text{ W}$$

The input power is obtained by adding this loss to the gap power, and is 38 850 W. Dividing the net output power of 36 194 W by this value gives the efficiency as 0.932 or 93.2%.

(e) When the slip is 1.0, the total impedance of the stator and rotor circuits is

$$\hat{Z} = 3.5 + j5.8 + 2.4 + j5.8 = 13.01 \,\underline{/63.04}\ \Omega$$

The rotor current is

$$\hat{I}'_2 = \frac{1270.2}{13.01 \,\underline{/63.04}} = 97.63 \,\underline{/-63.04} = 44.26 - j87.02 \text{ A}$$

The gap power is obtained as

$$P_g = 3 \times 97.63^2 \times 2.4$$
$$= 68\ 628 \text{ W}$$

Since the synchronous speed is 125.66 rad/s, the developed torque on starting is

$$T_d = \frac{68\ 628}{125.66} = 546.1 \text{ N·m}$$

(f) The condition for maximum developed torque is obtained using R_1 directly.

$$\frac{R'_2}{s} = |R_1 + jX_1 + jX'_2|$$
$$= |3.5 + j5.8 + j5.8| = 12.12\ \Omega$$

This gives the slip at maximum torque as 0.198, which is virtually the same as that obtained using the more accurate model in Example 7-2. The impedance of this circuit is

$$Z = |3.5 + j5.8 + 12.12 + j5.8|$$
$$= 19.46\ \Omega$$

The rotor current at maximum torque is

$$I'_2 = \frac{1270.2}{19.46} = 65.27 \text{ A}$$

The maximum gap power given by this model is

$$P_{gm} = 3 \times 65.27^2 \times 12.12 = 154\ 900 \text{ W}$$

and the maximum developed torque is

$$T_{dm} = \frac{154\ 900}{125.66} = 1233 \text{ N·m}$$

It should be evident that the errors introduced by the approximate circuit model are less than those introduced by neglecting the stator resistance in Example 7-2. As with the transformer, these errors are relatively greatest at light loads.

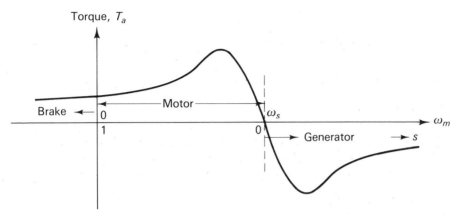

Figure 7-17   Complete Torque-Speed Characteristic.

In all the discussion so far it has been assumed that the slip is positive and significantly less than unity; that is, the speed is just less than the synchronous speed. However there is nothing in the expressions that have been derived to limit their validity to this limited range of slip or speed unless specifically noted as in Eq. (7-31). The complete torque-speed characteristic is shown in Fig. 7-17. The two additional parts of the characteristic derive from specific operating conditions.

The first condition is that when the speed is greater than synchronous speed. The slip is negative and therefore R_2'/s is also negative. This means that power is being injected into the rotor circuit from the shaft and the machine is operating as a generator. This is a very simple way of returning power into a large system.

The other condition is when the slip is greater than unity. This may be achieved by reversing the sequence of the source while the machine is motoring so that the rotating field reverses. In effect, the speed is now negative and the slip is a little less than 2.0 so that currents are relatively large. The torque developed is in the same direction as the rotating field and is therefore opposed to the motion. Since R_2'/s is positive, the power flow is still into the rotor and the machine is acting as a brake. Despite the large currents, the torques that are developed under this condition are not large, and since there is no regeneration, this is not a particularly attractive method of braking.

7.5 SQUIRREL-CAGE MACHINES

The description of the induction machine at the beginning of the chapter assumed that there would be a balanced polyphase winding on the rotor. Access to this winding is by means of slip rings. Normally a three-phase machine can be expected to have a three-phase rotor winding as well as the required three-phase winding on the stator, although it can have any number of phases. Since the slip under normal operating conditions is normally quite low, we can see from Eqs. (7-6) and (7-7) that the rotor voltage must be very low. It was discovered that there was no necessity to insulate the rotor conductors if there are no multiple turn coils. The "winding" thus can become a set of solid bars of solid copper connected at each end. One of the properties of such a system of conductors is that the voltages and currents induced in the rotor adjust themselves to the correct number of poles. The arrangement of conductors must have appeared to the developers of this arrangement as similar to a cage if the core is removed. Whatever the reason, it is now known as a *squirrel cage* which is extremely robust and requires no connections by means of slip rings. Current practice, at least for smaller motors, is to make the entire cage as a casting of aluminum. Figure 7-18 shows a rotor used in machines of four

Figure 7-18 Squirrel Cage Rotor (Photo courtesy of Canadian General Electric Co.).

or more poles. For high-speed applications the rotor diameter is restricted; Fig. 7-19 shows a sample with a corresponding stator in Fig. 7-20.

Since the conductors in a cage are solid, there is considerable skin effect, and the rotor resistance at low speeds is greater than that in the normal operating range. If trying to model such a machine over a wide range of speeds, it is necessary to modify the circuit model. There are several possibilities, but the simplest is based on an early form of squirrel

Figure 7-19   High-Speed Squirrel Cage Rotor (Photo courtesy of Canadian General Electric Co.).

Figure 7-20   Wound Stator for 2-Pole Squirrel Cage Induction Motor (Photo courtesy of Canadian General Electric Co.).

cage in which there were virtually two cages, one placed near the surface of the rotor and having a high resistance. This dominated the characteristics at low speed. The other was embedded more deeply within the rotor and had a low resistance. This dominated the characteristics in the normal operating range. This circuit model is shown in Fig. 7-21 where R_2' is the resistance of the outer cage and R_3' is that of the inner cage referred to the stator. Despite the fact that the shape of the bars is not simple, there have been many attempts to model the change in resistance directly. It is then more convenient to use Fig. 7-10 or Fig. 7-11 with values of rotor resistance and leakage reactance appropriate to the rotor frequency being considered.

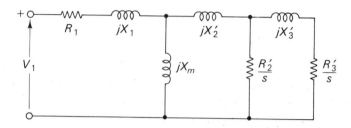

Figure 7-21   Circuit Model of Squirrel Cage Machine.

7.6 MEASUREMENT OF PARAMETERS

Since the model of an induction machine is essentially the same as that of
a transformer, it should not be surprising that the standard tests for an
induction machine are similar. However, one important difference is that
the primary and secondary resistances must be separated. It is not usually
necessary to separate the leakage reactances because the characteristics
are quite insensitive to the distribution of the total leakage reactance between
primary and secondary. It is therefore common to assume that the values,
referred to the primary, are equally divided.

7.6.1 DC Resistance Test

The dc resistance between pairs of stator terminals is measured and
provides an estimate of the primary resistance. If it is a wye connection
the value must be divided by two in order to obtain the resistance per
phase. For a delta connection the value is multiplied by 1.5 to get the
resistance per phase. In this case the temperature is most important, since
this measured value is to be subtracted from the equivalent resistance of
the complete motor obtained from the blocked rotor test considered in Sec.
7.6.3. The two tests should be done with winding temperatures as close
as possible. The two values must be referred to the same temperature
using the temperature coefficient of resistance for the winding.

7.6.2 No-Load Test

A no-load test that corresponds exactly with that of the transformer
would require that the rotor be rotated at exactly synchronous speed. Since
this is normally not possible, the motor is run on no-load so that measurements
of current, voltage, and power may be obtained. The load in this test
consists only of losses as follows:

(a) hysteresis and eddy current losses in the stator and rotor cores,

(b) windage and friction,

(c) copper loss in the windings.

If the current on no-load is very much less than when operating at rated load, the no-load copper loss may be negligible. However, the presence of the air gap in the magnetic system requires a higher magnetizing current than that in a transformer, and the no-load copper loss may not be negligible. Whether or not this is the case can be determined only after the data from the blocked rotor test is available.

The power factor on no-load is usually quite low, and the reactive component can be interpreted as being essentially the magnetizing current. The in-phase component of the current *cannot* be used to determine any resistance, and this is why it is usually convenient to consider that the no-load power is the core, windage, and friction loss after due allowance has been made for the copper loss in the conductors.

With the approximations noted above, the reactance is calculated as follows, using phase currents and voltages in Eq. (5-24).

$$X_{n-l} = \frac{V_{n-l}}{I_{n-l} \sin \beta_{n-l}}$$

where V_{n-l} is the stator phase voltage, I_{n-l} is the magnitude of the no-load current, and β_{n-l} is the no-load phase angle in Fig. 7-9.

If the approximate circuit model is acceptable, this may be interpreted directly as the magnetizing reactance. For the more accurate model of Fig. 7-10 it must be taken as a series circuit:

$$X_{n-l} = \frac{V_{n-l} \sin \beta_{n-l}}{I_{n-l}} = X_1 + X_m \tag{7-32}$$

7.6.3 Blocked Rotor Test

As its name implies, the rotor is restrained by blocking it so that this test corresponds exactly with the short circuit test of a transformer. The measurements of current, voltage, and power are used to get the equivalent impedance directly if the approximate circuit model of Fig. 7-11 is being used. When the parameters of the more accurate model of Fig. 7-10 are required, it is necessary to combine the data from all three tests. The current for the blocked rotor test is usually set to the rated value, the voltage required being much less than the rated value. Inevitably, the temperature rises during this test. If the resistance calculated is to be combined with that obtained from the dc measurement of the stator resistance, these values must be referred to the same temperature, preferably that during normal operation.

Referring to Fig. 7-10, but noting that $s = 1.0$ for this test, Eq. (7-15) can be expanded to give

$$R_f = \frac{R_2' X_m^2}{R_2'^2 + (X_2' + X_m)^2}$$

and

$$X_f = \frac{X_m[R_2'^2 + X_2'(X_2' + X_m)]}{R_2'^2 + (X_2' + X_m)^2}$$

Since R_2' is very much less than $(X_2' + X_m)$, there is little error in neglecting the $R_2'^2$ terms. Thus

$$R_f = R_2'\left[\frac{X_m^2}{(X_2' + X_m)^2}\right]$$

$$= k^2 R_2'$$

(7-33)

and

$$X_f = kX_2'$$

(7-34)

where

$$k = \frac{X_m}{X_2' + X_m}$$

(7-35)

The equivalent resistance and reactance at zero speed are obtained from the measured values of current, voltage, and power. The calculations are similar to the interpretation of the short circuit test of a transformer given in Eqs. (5-22) to (5-28).

$$R_{bl-r} = R_1 + R_f = R_1 + k^2 R_2'$$

(7-36)

and

$$X_{bl-r} = X_1 + X_f = X_1 + kX_2'$$

(7-37)

If we now assume that $X_1 = X_2'$, Eqs. (7-32) to (7-34) can be solved for the constant k.

$$X_{bl-r} = X_1(1 + k)$$

$$X_1 = X_{n-l}(1 - k)$$

(7-38)

from which

$$X_{bl-r} = X_{n-l}(1 - k^2)$$

and

$$k^2 = 1 - \frac{X_{bl-r}}{X_{n-l}}$$

(7-39)

Having found a close approximation to the value of k, all the parameter values may now conveniently be found.

One major complication with squirrel-cage motors is the frequency dependence of the rotor resistance and leakage reactance. It is very difficult to measure the parameter values of Fig. 7-21, but it is possible to use Fig. 7-10 over a restricted range of speed. It is necessary to use a low frequency corresponding to the slip frequency under normal load conditions. Clearly, the values of reactance must be adjusted to the operating frequency before being used in the above expressions for the interpretation of test data.

Example 7-4

A three-phase, 220-V, 60-Hz, four-pole wound rotor induction motor was subjected to blocked rotor and no-load tests. The current, voltage, and power when the rotor was blocked were 14.13 A, 26.94 V, and 288.5 W, respectively. With no load applied to the shaft, the corresponding measurements were 8.73 A, 220 V, and 659 W, the speed being 1782 r/min. The dc resistance measured between a pair of stator terminals was 0.4 Ω.

Determine the parameters of its circuit model.

Solution. The internal connection of the motor is not known. This does not pose any real problem, since we are free to determine the parameters of either the equivalent wye or delta connection. In this case the equivalent wye will be used.

The dc resistance between a pair of terminals in a wye connection is twice the resistance of one phase, since the terminal of the other phase is not connected. Thus we may write

$$R_1 = \frac{0.4}{2} = 0.2 \ \Omega$$

From the blocked rotor test we obtain the magnitude of the phase impedance as

$$Z_{bl-r} = \frac{26.94}{\sqrt{3} \times 14.13} = 1.1007 \ \Omega$$

The power factor is obtained from the measurements of current, voltage, and power as

$$\cos \beta = \frac{288.5}{\sqrt{3} \times 26.94 \times 14.13} = 0.4375$$

and

$$\sin \beta = 0.8992$$

Thus the blocked rotor impedance is

$$\hat{Z}_{bl-r} = R_{bl-r} + jX_{bl-r} = 0.4816 + j0.9897 \ \Omega$$

The no-load test data is used only to determine the approximate value of the reactance, since the power required for the mechanical losses is unknown. The power factor is

$$\cos \beta = \frac{659}{\sqrt{3} \times 220 \times 8.73} = 0.1981$$

and

$$\sin \beta = 0.980$$

The no-load reactance is therefore

$$X_{n-l} = \frac{220}{\sqrt{3} \times 8.73} 0.980 = 14.26 \ \Omega$$

We are now in a position to apply Eq. (7-39) so that

$$k = \sqrt{\left(1 - \frac{X_{bl-r}}{X_{n-l}}\right)} = \sqrt{\left(1 - \frac{0.9897}{14.26}\right)} = 0.9647$$

The leakage reactance of the stator can now be determined, using Eq. (7-38), as

$$X_1 = 14.26 \times (1 - 0.9647) = 0.5034 \; \Omega$$

We now recall that Eq. (7-38) was based on the premise that $X_1 = X_2'$. Thus the value of the rotor leakage reactance, referred to the stator, is also $0.5034 \; \Omega$. The rotor resistance, R_2', is obtained from Eq. (7-36) as

$$R_2' = \frac{R_{bl-r} - R_1}{k^2} = 0.3025 \; \Omega$$

To complete the model, we rearrange Eq. (7-35) to get

$$X_m = \frac{k X_2'}{1 - k} = 13.76 \; \Omega$$

7.7 CONTROL OF INDUCTION MOTORS

Usually the rotor resistance is relatively low and the slip in the normal operating range is quite small, generally less than 5%. As a result, it is difficult to vary the speed of an induction motor connected to a constant frequency source.

For many years the most common method required a wound rotor motor and a three-phase resistance connected to its slip rings. This provides a method of increasing the rotor resistance, similar to connecting resistance in series with the armature of a dc motor. This is illustrated in Fig. 7-22, where three torque-speed characteristics corresponding to three different values of R_2' are shown. This method is inefficient, and when the rotor resistance has been increased, the variation of speed due to changes in load torque also increases. One point in its favor is that the starting torque can be increased while at the same time limiting the starting current.

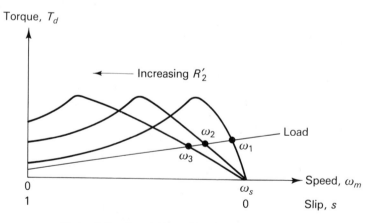

Figure 7-22   Speed Control by Rotor Resistance.

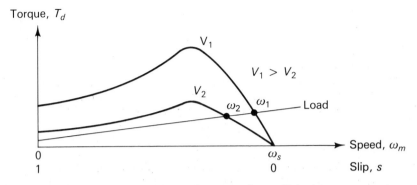

Figure 7-23   Speed Control by Source Voltage.

Referring to Eq. (7-28), it can be seen that the developed torque is proportional to the square of the source voltage. A reduction in terminal voltage will produce a reduction in speed. Figure 7-23 shows torque-speed characteristics for approximately 70% and 100% of rated voltage. At the lower voltage the slip must increase, but there is a possibility that the current will increase despite the reduced voltage; this method must be used with caution.

The most effective way of changing the speed of an induction motor is to change the frequency. However, it is necessary to change the voltage at the same time, very often in direct proportion to the frequency. The need for this can be seen by referring to Eq. (3-23). When the ratio of induced voltage to frequency is kept constant, the flux density in the various parts of the core can be maintained at a constant level. This is usually called constant volts per hertz control. When this is done the maximum torque is independent of frequency, and the slip at maximum torque increases as the frequency is decreased. Three characteristics are shown in Fig. 7-24, where it is evident that as long as the slip at any operating frequency is low, the speed is approximately proportional to the frequency.

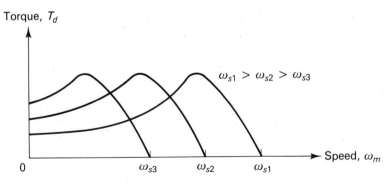

Figure 7-24   Control by Frequency.

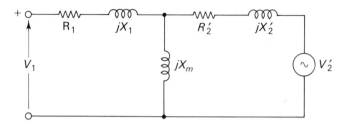

Figure 7-25 Slip Energy Recovery.

The variable frequency source usually involves power electronics circuits. The basic principles employed are described in Sect. 7.7.1. However, these circuits generally produce nonsinusoidal waveforms, and the model of the motor developed in this chapter is not adequate. In practice these drive systems require extensive control because of severe problems with the dynamic characteristics of the induction motor when subjected to this form of excitation. Such problems are outside the scope of this text.

If the additional resistance in a wound rotor motor is replaced by a polyphase source having the exact slip frequency, it is possible to recover the energy which would otherwise be wasted, as noted above. In the circuit shown in Fig. 7-25 it is possible to transfer power, provided the two sources have the same frequency. Thus, there must be an arrangement to ensure that the frequency of the source connected to the slip rings is automatically constrained to be the same as that of the source. The first of such slip-energy-recovery schemes to be developed employed a commutator as the combined frequency changer and slip rings. These ac commutator machines were expensive to maintain and did not find extensive application. With the availability of thyristor circuits that can achieve the same characteristics, there is some interest in such machines, but the improvements achieved with squirrel-cage motors fed by a variable-frequency source have inhibited development of slip-energy-recovery systems.

It is possible to construct a winding which by means of simple switching can change its effective number of poles from one value to another. This can be of interest if only change of speed to a limited number of discrete values is desired. Unfortunately, a discussion of such motors requires extensive knowledge of winding design. The design invokes concepts of amplitude modulation and such motors are known as *pole amplitude modulated* (PAM) induction motors, although the actual layout of the winding is closer to pulse-width modulation.

7.7.1 DC–AC Converters

The control of the speed of an induction motor by varying the frequency requires a suitable source. When discussing this form of control, it was noted that a strategy of constant volts per hertz is required; thus the source

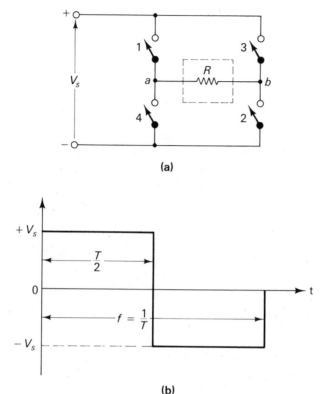

Figure 7-26 Single-phase Inverter.

must provide a variable-frequency voltage that is proportional to the frequency. The amplitude of the voltage waveform is normally varied by means of a controlled rectifier such as those described in Sec. 6-7, although a chopper may be used. The resulting dc is inverted to give the necessary ac input to the motor. An *inverter* is essentially a circuit having switches that are switched on and off to provide a square waveform. The principle of operation is explained by referring to Fig. 7-26a, where the semiconductor devices have been shown as simple switches. When switches 1 and 2 are closed, the voltage across the ideal resistive load is $+V_s$, but when switches 3 and 4 are closed the voltage across the load is $-V_s$. If these pairs of switches are alternately closed and opened, the voltage across the load is that shown in Fig. 7-26b.

As long as the switching is performed periodically with equal times for the two states, the output is a standard square wave. The harmonic series representing this waveform is

$$v_{ab} = \sum_{n=1} \frac{4V_s}{n\pi} \sin 2n\pi ft \tag{7-40}$$

so that the amplitude of the fundamental is

$$V_1 = \frac{4V_s}{\pi} \tag{7-41}$$

The semiconductor devices used as switches can be forced commutated thyristors, gate turn-off thyristors, or transistors. The basic operation of the switch is similar to that of the dc chopper.

The harmonic series given by Eq. (7-40) indicates that there is significant harmonic content in the voltage waveform. The resulting harmonic currents cause increased losses and hence reduced efficiency of the motor. However, in many applications this reduction in efficiency of the motor is very much less than the increase in efficiency gained by using this form of control.

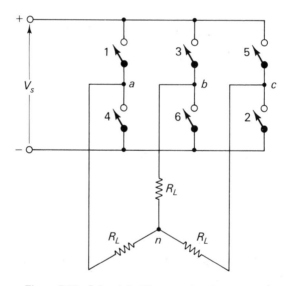

Figure 7-27 Schematic Diagram of 3-phase Inverter.

For control of a three-phase motor a three-phase inverter is used; the basic schematic diagram is shown in Fig. 7-27, where the switches are controlled in much the same manner as those of the single-phase inverter. The resulting waveforms are shown in Fig. 7-28. Fourier analysis of these waveforms gives the following expressions for the three line voltages.

$$v_{ab} = \sum_{n=1} \frac{4V_s}{n\pi} \cos \frac{n\pi}{6} \sin n\left(\omega t + \frac{\pi}{6}\right)$$

$$v_{bc} = \sum_{n=1} \frac{4V_s}{n\pi} \cos \frac{n\pi}{6} \sin n\left(\omega t - \frac{\pi}{2}\right) \tag{7-42}$$

$$v_{ca} = \sum_{n=1} \frac{4V_s}{n\pi} \cos \frac{n\pi}{6} \sin n\left(\omega t + \frac{5\pi}{6}\right)$$

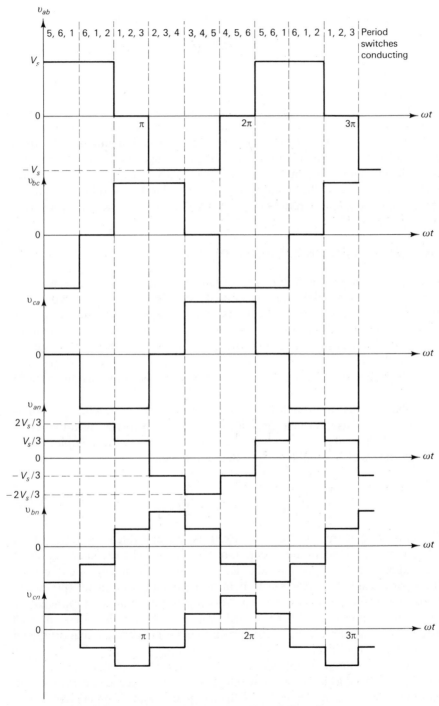

Figure 7-28 Voltage Waveforms in a 3-phase Inverter.

The amplitude of the fundamental of these waveforms is

$$V_1 = \frac{2\sqrt{3}}{\pi} V_s \qquad (7\text{-}43)$$

and the phase angle between each pair can be seen to be $2\pi/3$ or $120°$, thus forming a balanced three-phase set.

To complete this discussion, it may be observed that an alternative approach to controlling the amplitude of the ac output could be to incorporate a chopping action within the inverter. Different techniques may be used to minimize the harmonics, thus producing a voltage at the motor terminals more closely approximating a sinusoid. The motor may be connected either in delta or wye, and the main complication is due to its inductive nature. This requires additional circuitry to enable the switches to commutate properly.

7.8 SINGLE-PHASE INDUCTION MOTORS

Single-phase induction motors are usually used in small sizes where the rating is less than a kilowatt (fractional horsepower motors). However, there are some domestic applications, such as air-conditioning, requiring ratings up to the order of 10 kW. Normally, a single-phase induction motor has a squirrel-cage rotor. It may be noted in passing that a three-phase motor will continue to operate as a single-phase motor if one of the lines becomes disconnected. The result is increased noise and current, both of which may be difficult to detect.

The simplest approach to the modeling of the single-phase motor is to resolve the pulsating field produced by the single current into two components. With only one winding, Eq. (7-1) represents the complete mmf distribution, which is split into its two components.

$$\begin{aligned}
\mathscr{F}_x &= N_x I_x \cos\theta \cos\omega t \\
&= \frac{N_x I_x}{2} \cos(\theta - \omega t) + \frac{N_x I_x}{2} \cos(\theta + \omega t)
\end{aligned} \qquad (7\text{-}44)$$

The first component can be recognized as the equivalent of Eq. (7-3), a field pattern traveling in the direction of positive ω. The second component is a field pattern of the same amplitude, but traveling in the opposite direction. The squirrel-cage rotor reacts with both of these component fields, producing a positive torque in response to the forward-traveling field and a negative torque in response to the backward-traveling field. Other than at zero speed, the torque produced by the forward field is greater than that produced by the backward field.

All the terminology introduced for the three-phase motor may be used. The slip is defined as in Eq. (7-8), with the synchronous speed being associated with the forward field. The value of the slip under normal running conditions

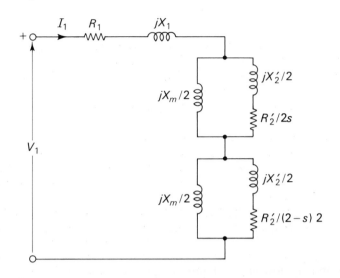

Figure 7-29   Circuit Model of Single-phase Induction Motor.

is typically low, although not as low as that of a three-phase motor. With this value of slip (relative to the forward wave), the slip relative to the backward wave is

$$s_b = 2 - s \tag{7-45}$$

Since s is small, the value of s_b is normally only slightly less than two.

The circuit model is obtained by modifying that of the three-phase motor and noting that the amplitude of the forward and backward fields is half of that produced by the stator winding. The rotor circuits shown in Fig. 7-29 therefore have all their parameter values multiplied by this factor. We may note that at $s = 1.0$ the values in the forward rotor circuit equal those in the backward circuit, and the complete circuit is the same as that of one phase of a three-phase motor. Little can be done to simplify this circuit, and the following summary is based on the solution using the equivalent impedance of the rotor and magnetizing branches described in Sec. 7.4.1. This produces the simple series circuit shown in Fig. 7-30.

If we let $\hat{Z}_f$ be the equivalent of the forward circuit, corresponding to Eq. (7-15), then $\hat{Z}_b$ is the equivalent of the backward circuit, given by

$$\hat{Z}_b = \frac{1}{2} \left[\frac{\left(\dfrac{R_2'}{2 - s} + jX_2' \right)(jX_m)}{\dfrac{R_2'}{2 - s} + j(X_2' + X_m)} \right] \tag{7-46}$$

$$= R_b + jX_b$$

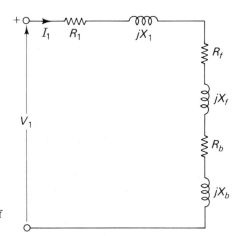

Figure 7-30 Series Equivalent Circuit of
Single-phase Induction Motor.

The solution may now proceed quite simply by calculating the input current
to the motor, I_1. The gap power associated with the forward field is

$$P_{gf} = I_1^2 R_f \tag{7-47}$$

The gap power associated with the backward field is

$$P_{gb} = I_1^2 R_b \tag{7-48}$$

These two components of the gap power produce corresponding torques.
The developed torque is the difference between these two torques, which
are obtained using Eq. (7-14).

$$T_d = T_{df} - T_{db} \tag{7-49}$$

$$= \frac{1}{\omega_s}(P_{gf} - P_{gb})$$

The gross mechanical power is obtained by subtracting the rotor copper
losses from the net gap power. The copper losses are

$$P_{2f} = sP_{gf}$$

$$P_{2b} = (2 - s)P_{gb}$$

Thus

$$P_m = P_{gf} - P_{gb} - P_{2f} - P_{2b} \tag{7-50}$$

If the rotational losses are known, the net output or shaft power can now
be determined.

$$P_{sh} = P_m - P_{n-l} \tag{7-51}$$

Example 7-5

A one-phase, 110-V, 60-Hz, four-pole induction motor has the following parameter
values:

$$R_1 = 1.5 \ \Omega \qquad X_1 = 2.5 \ \Omega$$
$$R_2' = 3.5 \ \Omega \qquad X_2' = 2.5 \ \Omega$$
$$X_m = 50.0 \ \Omega$$

The no-load losses are 20 W. Determine the input current, input power, power factor, developed torque, and efficiency when operating at a speed of 1746 r/min.

Solution. The data corresponds to Fig. 7-29 so that after determining the value of the slip, the values for the circuit of Fig. 7-30 can be obtained. Since this is a 60-Hz, four-pole motor, the synchronous speed is 1800 r/min and the slip is

$$s = \frac{1800 - 1746}{1800} = 0.03$$

$$\hat{Z}_f = 0.5 \left\{ \frac{\left[\left(\dfrac{3.5}{0.03} \right) + j2.5 \right] \times j50.0}{\left(\dfrac{3.5}{0.03} \right) + j(2.5 + 50.0)} \right\}$$

$$= 8.91 + j20.991 \ \Omega$$

$$\hat{Z}_b = 0.5 \left\{ \frac{\left[\left(\dfrac{3.5}{1.97} \right) + j2.5 \right] \times j50.0}{\left(\dfrac{3.5}{1.97} \right) + j(2.5 + 50.0)} \right\}$$

$$= 0.805 + j1.218 \ \Omega$$

$$\hat{Z}_1 = 1.5 + j2.5 \ \Omega$$

The total impedance is

$$\hat{Z} = \hat{Z}_1 + \hat{Z}_f + \hat{Z}_b$$
$$= 11.215 + j24.709 \ \Omega$$

The input current is

$$\hat{I}_1 = \frac{110}{11.215 + j24.709} = 4.054 \ \angle -65.59 \ \text{A}$$

That is, the input current is 4.054 A and the power factor is the cosine of 65.59°, namely 0.413 (lagging). The input power is

$$P_{in} = VI \cos \beta = 110 \times 4.054 \times 0.413 = 184.2 \ \text{W}$$

The components of the gap power are

$$P_{gf} = 4.054^2 \times 8.91 = 146.44 \ \text{W}$$
$$P_{gb} = 4.054^2 \times 0.805 = 13.23 \ \text{W}$$

Since the synchronous speed is 188.5 rad/s, the developed torque is

$$T_d = \frac{146.44 - 13.23}{188.5} = 0.707 \ \text{N·m}$$

The gap power can be obtained by multiplying the developed torque by the angular speed of the shaft. This gives

$$P_g = 0.707 \times 188.5 \times 0.97 = 129.1 \text{ W}$$

This is, of course, simply the difference between P_{gf} and P_{gb}, and could have been obtained directly by subtraction.

To determine the efficiency, the rotor copper losses must now be calculated. These are

$$P_{2f} = 146.44 \times .03 = 4.39 \text{ W}$$

$$P_{2b} = 13.23 \times 1.97 = 26.06 \text{ W}$$

$$P_m = 146.44 - 13.23 - 4.39 - 26.06 = 102.76 \text{ W}$$

$$P_{sh} = P_m - P_{n-l} = 102.76 - 20 = 82.76 \text{ W}$$

Note that the input power can also be obtained by adding the stator copper loss to the total gap power.

$$P_{in} = 146.44 + 13.23 + 4.054^2 \times 1.5 = 184.32 \text{ W}$$

This compares with the value of 184.2 W, determined using the input voltage and current relations, the small difference being due to round-off errors in calculation. Note particularly that P_{gb} is added to P_{gf}, since both components are supplied from the stator. The efficiency is therefore $82.76/184.2 = 0.449$ or 44.9%. Although it is not possible to generalize on the basis of one calculation, it should be noted that the efficiency of a single-phase induction motor is usually less than that of the corresponding three-phase motor.

In the derivation of the circuit model it was assumed that the forward and backward field components had equal amplitudes. For the three-phase induction motor, the voltage across the magnetizing branch is proportional to the amplitude of the revolving field. In this case the voltages across Z_f and Z_b are proportional to the amplitudes of the components of the field. The only speed at which these two impedances are equal, resulting in equal amplitudes, is the synchronous speed. At all other speeds in the normal motoring region of the complete characteristic the amplitude of the forward field is the greater of the two. This has the merit of improving the performance, but the single-phase motor is nevertheless subject to pulsating torques which cannot be avoided.

7.8.1 Starting Single-Phase Induction Motors

At the moment of starting the slip is 1.0 and R_f equals R_b so that the developed torque, T_d, is zero. It is thus necessary to find some alternate means to develop the required starting torque. There are three principal methods of doing so:

1. split-phase motor,
2. capacitor-start motor,
3. shaded pole motor.

Each of these methods consists of providing an additional component of revolving field such that there is a quasi two-phase effect. In the case of the split phase and capacitor-start motors, this is done by adding a second winding similar to that in a two-phase motor. For the *split phase* motor, the ratio of reactance to resistance is different in the two windings so that there is a (time) phase angle between the two currents. Although this does not produce a balanced two-phase system such as that considered at the beginning of this chapter, it is sufficient to produce a small starting torque. If a greater starting torque is required, a *capacitor start* motor gets closer to the ideal two-phase situation by inserting a capacitor in series with the starting winding. In both cases a centrifugal switch is operated after the rotor is accelerated to approximately three quarters of rated speed, thus disconnecting the starting winding and allowing the motor to operate in its normal mode, described in Sec. 7.8.

The *shaded-pole* motor is generally used in very small sizes, a common example being the drive motor in inexpensive record players. In this case there is only one winding, and the stator has salient poles as shown in Fig. 7-31. Around part of each of these poles is placed a short-circuiting ring. The action of the currents induced in these short circuits is to cause a time lag in the magnetic field crossing that part of the gap. Since there are now two components of field having a (time) phase shift, the total field has some of the properties of a two-phase motor, and the result is a net torque at zero speed. There is no attempt to open-circuit the *shading ring* during normal operation.

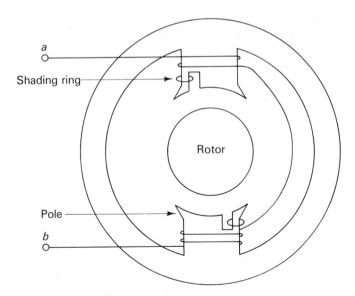

Figure 7-31 Shaded Pole Motor.

PROBLEMS

7-1. A three-phase, 220-V, 60-Hz, six-pole, wye-connected, wound rotor induction motor has 170 V between slip rings at standstill when a balanced 220-V, three-phase source is connected to its stator terminals. The rotor is coupled directly to a dc motor whose speed can be changed. Assuming that there is a sinusoidal field distribution, determine the magnitude and frequency of the voltage between slip rings when the rotor is driven at the following speeds.
  (a) 1150 r/min in the same direction as the rotating field,
  (b) 1150 r/min in the opposite direction to the rotating field,
  (c) 1200 r/min in the same direction as the rotating field,
  (d) 1500 r/min in the same direction as the rotating field,
  (e) 3600 r/min in the opposite direction to the rotating field.

7-2. A three-phase, 460-V, 60-Hz, four-pole, wye-connected, wound rotor induction motor has 235 V between slip rings at standstill when a balanced 460-V, three-phase source is connected to its stator terminals. The rotor is coupled directly to a dc motor whose speed can be changed. Assuming that there is a sinusoidal field distribution, determine the magnitude and frequency of the voltage between slip rings when the rotor is driven at the following speeds.
  (a) 1750 r/min in the same direction as the rotating field,
  (b) 1750 r/min in the opposite direction to the rotating field,
  (c) 1800 r/min in the same direction as the rotating field,
  (d) 1800 r/min in the opposite direction to the rotating field,
  (e) 2000 r/min in the same direction as the rotating field.

7-3. A three-phase, 460-V, 60-Hz, eight-pole, wye-connected, wound rotor induction motor has a stator resistance of 0.45 Ω per phase. When run on no-load the measured quantities are:

$$\text{Line voltage} \;=\; 460 \text{ V}$$
$$\text{Line current} \;=\; 7 \text{ A}$$
$$\text{Input power} \;=\; 2000 \text{ W}$$

The blocked rotor test measurements at 60 Hz were:

$$\text{Line voltage} \;=\; 80 \text{ V}$$
$$\text{Input power} \;=\; 1650 \text{ W}$$
$$\text{Line current} \;=\; 25 \text{ A}$$

Obtain the exact equivalent circuit of the motor.

7-4. A three-phase, 208-V, 60-Hz, six-pole, wye-connected, wound rotor induction motor has a stator resistance of 0.25 Ω per phase. When run on no-load the measured quantities are:

$$\text{Line voltage} \;=\; 208 \text{ V}$$
$$\text{Line current} \;=\; 3 \text{ A}$$
$$\text{Input power} \;=\; 500 \text{ W}$$
$$\text{Windage and friction loss} \;=\; 300 \text{ W}$$

The blocked rotor test measurements at 60 Hz were:

$$\text{Line voltage} = 32.5 \text{ V}$$
$$\text{Input power} = 350 \text{ W}$$
$$\text{Line current} = 12 \text{ A}$$

(a) Obtain the approximate equivalent circuit.

(b) If the motor runs at 1170 r/min, use the approximate circuit model to determine
   (1) the input current,
   (2) the developed torque,
   (3) the output power,
   (4) the output torque,
   (5) the efficiency.

(c) Determine the slip at which maximum torque is developed and calculate its value.

7-5. A three-phase, 460-V, 60-Hz, eight-pole, wye-connected, wound rotor induction motor has a stator resistance of 0.2 Ω per phase. When run on no-load the measured quantities are:

$$\text{Line voltage} = 460 \text{ V}$$
$$\text{Line current} = 4 \text{ A}$$
$$\text{Input power} = 750 \text{ W}$$
$$\text{Windage and friction loss} = 500 \text{ W}$$

The blocked rotor test measurements at 60 Hz were:

$$\text{Line voltage} = 43.5 \text{ V}$$
$$\text{Input power} = 600 \text{ W}$$
$$\text{Line current} = 20 \text{ A}$$

(a) Obtain the exact equivalent circuit.

(b) If the motor runs at 880 r/min, use the exact circuit model to determine
   (1) the input current,
   (2) the developed torque,
   (3) the output power,
   (4) the output torque,
   (5) the efficiency.

(c) Determine the slip at which maximum torque is developed and calculate its value.

7-6. A three-phase, 440-V, 60-Hz, six-pole, wye-connected, wound rotor induction motor has the following parameters per phase:

$$\text{Stator resistance} = 0.5 \text{ Ω}$$
$$\text{Stator leakage inductance} = 0.003 \text{ H}$$
$$\text{Rotor leakage inductance} = 0.003 \text{ H}$$
$$\text{Magnetizing inductance} = 0.1 \text{ H}$$
$$\text{Rotor resistance} = 0.3 \text{ Ω}$$
$$\text{Turns ratio} = 1:1$$

Rotational losses negligible

The stator is connected to a source of rated voltage and frequency. Use the approximate equivalent circuit to determine

(a) the standstill torque with short-circuited rotor,

(b) the maximum torque that can be produced with the rotor short-circuited,

(c) the speed at which this maximum torque is developed.

7-7. A three-phase, 550-V, 60-Hz, eight-pole, wye-connected, wound rotor induction motor has the following parameters per phase:

Stator resistance	=	$0.6\ \Omega$
Stator leakage inductance	=	0.004 H
Rotor leakage inductance	=	0.004 H
Magnetizing inductance	=	0.125 H
Rotor resistance	=	$0.4\ \Omega$
Turns ratio	=	$1{:}1$
Rotational losses negligible		

The stator is connected to a source of rated voltage and frequency. Use the exact equivalent circuit to determine

(a) the standstill torque with short-circuited rotor,

(b) the maximum torque that can be produced with the rotor short-circuited,

(c) the speed at which this maximum torque is developed.

7-8. The resistance measured between each pair of slip rings of a three-phase, 60-Hz, ten-pole, wye-connected, wound rotor induction motor is $0.042\ \Omega$. The slip when delivering rated load with the slip rings short-circuited is 0.03, and it may be assumed that developed torque is proportional to slip in the region between no-load and rated load.

The motor drives a mechanical system that requires 200 kW at the rated speed of the motor. This mechanical system has a torque-speed characteristic proportional to the square of the speed. A wye-connected resistance is to be connected to the slip rings so that the speed when driving this system is 400 r/min.

Determine the resistance per phase required.

7-9. The resistance measured between each pair of slip rings of a three-phase, 60-Hz, eight-pole, delta-connected, wound rotor induction motor is $0.096\ \Omega$. The slip, when delivering rated load with the slip rings short-circuited, is 0.025, and it may be assumed that developed torque is proportional to slip in the region between no-load and rated load.

The motor drives a mechanical system which requires 150 kW at the rated speed of the motor. This mechanical system has a torque-speed characteristic proportional to the speed. A delta-connected resistance is to be connected to the slip rings so that the speed when driving this system is 450 r/min.

Determine the resistance per phase required.

7-10. A three-phase induction motor runs at a slip of 0.04 when operating at rated load. When started direct on line the input current is eight times the rated value. If the rotor resistance is taken as independent of rotor frequency, the magnetizing current, the mechanical and stray losses are neglected.

(a) Determine the starting torque as a ratio of the rated torque.

(b) If the stator resistance is negligible, determine the maximum torque as a ratio of rated torque, and the corresponding slip.

7-11. A three-phase, 460-V, six-pole, wye-connected, wound rotor induction motor runs at a slip of 0.08 when the developed torque is twice the value at rated load. The input current for this condition is 2.3 times the rated value. The rotor is wye-connected with a resistance of 0.05 Ω per phase. Determine the resistance per phase of a set of wye-connected resistors connected to the slip rings, which will limit the starting current at rated voltage and frequency to 2.3 times the rated value. Determine also the ratio of the starting torque to the rated value.

7-12. A three-phase, 2200-V, 60-Hz, eight-pole, wye-connected, squirrel-cage induction motor is used to drive a pump. At standstill the rotor leakage inductance and resistance are respectively 0.5 and 1.6 times their normal steady-state values when the slip is low. The parameters in ohms per phase for normal operation are

$$R_1 = 1.17 \qquad X_1 = 2.34$$
$$R_2' = 1.75 \qquad X_2' = 2.34$$
$$X_m = 130$$

The steady-state speed is 872 r/min.

Using the approximate equivalent circuit, determine

(a) the steady-state input current,

(b) the magnitude of the voltage to be used on starting if the magnitude of the input current on starting is to be limited to twice the steady-state current.

7-13. A three-phase, 4000-V, 60-Hz, 12-pole, wye-connected, squirrel-cage induction motor is used to drive a pump. At standstill the rotor leakage inductance and resistance are respectively 0.7 and 1.8 times their normal steady-state values when the slip is low. The parameters in ohms per phase for normal operation are

$$R_1 = 1.84 \qquad X_1 = 3.13$$
$$R_2' = 2.05 \qquad X_2' = 3.13$$
$$X_m = 190$$

The steady-state speed is 591 r/min.

Using the exact equivalent circuit, determine

(a) the steady-state input current,

(b) the magnitude of the voltage to be used on starting if the magnitude of the input current on starting is to be limited to twice the steady-state current.

7-14. A three-phase, 550-V, 60-Hz, six-pole, wye-connected, squirrel-cage induction motor has the following parameter values in ohms per phase.

$$R_1 = 1.3 \qquad X_1 = 4.5$$
$$R_2' = 1.8 \qquad X_2' = 4.5 \text{ at frequencies less than 5 Hz}$$
$$R_2' = 3.9 \qquad X_2' = 3.5 \text{ at 60 Hz}$$
$$X_m = 250$$

Determine the input current and developed torque when the motor runs at a speed of 1175 r/min.

For starting, the applied voltage is to be reduced so that the starting current is limited to twice the value found above. Determine the voltage required and the resulting starting torque.

7-15. A three-phase, 460-V, 60-Hz, two-pole, wye-connected, squirrel-cage induction motor has the following parameter values in ohms per phase.

$$R_1 = 1.1 \qquad X_1 = 3.5$$
$$R_2' = 1.5 \qquad X_2' = 3.8 \text{ at frequencies less than 5 Hz}$$
$$R_2' = 3.6 \qquad X_2' = 3.2 \text{ at 60 Hz}$$
$$X_m = 350$$

Determine the input current and developed torque when the motor runs at a speed of 3500 r/min.

For starting, the applied voltage is to be reduced so that the starting current is limited to twice the value found above. Determine the voltage required and the resulting starting torque.

7-16. A one-phase, 115-V, 60-Hz, four-pole induction motor has the following resistances and reactances in ohms.

$$R_1 = 0.5 \qquad X_1 = 0.4$$
$$R_2' = 0.25 \qquad X_2' = 0.4$$
$$X_m = 35$$

Determine the input current and developed torque when the motor runs at a speed of 1730 r/min.

7-17. A one-phase, 120-V, 60-Hz, two-pole induction motor has the following resistances and reactances in ohms.

$$R_1 = 0.6 \qquad X_1 = 0.5$$
$$R_2' = 0.28 \qquad X_2' = 0.5$$
$$X_m = 40$$

Determine the input current and developed torque when the motor runs at a speed of 3500 r/min.

7-18. A three-phase, 440-V, 60-Hz, six-pole, delta-connected, wound rotor induction motor has the following parameters per phase:

Stator resistance	=	1.5 Ω
Stator leakage inductance	=	0.008 H
Rotor leakage inductance	=	0.008 H
Magnetizing inductance	=	0.3 H
Rotor resistance	=	0.8 Ω
Turns ratio	=	1:1
Rotational losses negligible		

The stator is connected to a source of rated voltage and frequency. Use the approximate circuit model to calculate and plot the characteristics speed/developed torque and slip/developed torque, and hence determine the maximum developed torque and the speed at which this occurs.

7-19. The motor whose parameter values are given in Prob. 7-18 is to be used with a variable-frequency source and operated with constant volts per hertz. Use the approximate circuit model to calculate and plot the developed torque against speed characteristic for frequencies of

(a) 60 Hz,

(b) 45 Hz,

(c) 30 Hz.

(d) For each of these operating frequencies, determine the speed at which the developed torque is maximum.

7-20. The motor whose parameter values are given in Prob. 7-18 is to be used with a variable voltage source and operated with constant frequency. Use the approximate circuit model to calculate and plot the developed torque against speed characteristic for terminal voltages that are

(a) 100%,

(b) 75%,

(c) 50% of rated value.

(d) For each of these operating voltages, determine the speed at which the developed torque is maximum.

8

SYNCHRONOUS MACHINES

8.1 INTRODUCTION

In this chapter we shall study machines having the property that the speed of the rotor is equal to or is a submultiple of the frequency of the source. This was observed in Example 4-3 and resulted from the fact that there was only one speed at which the average value of the developed torque was some value other than zero. We shall consider synchronous machines having a polyphase armature winding, usually on the stator, and a field winding carrying direct current on the rotor.

The windings on synchronous motors and generators may have multiple poles, just like induction machines, with the result that the ratio of the frequency of the source to the speed is always some simple integer number. In practice, most utilities operate their systems with very little variation in the frequency, so that a synchronous motor can be expected to operate with very little change in speed. However, it must be noted that just as the frequency of even the best of power systems is not absolutely constant, the speed of a synchronous motor is not absolutely constant. The synchronous generator or alternator is widely used in power systems with ratings up to 750 MW or more.

The excitation is normally provided by a field winding, which is usually placed on the rotor. Sometimes the coils are mounted on pole pieces that project; such a machine is said to have *salient poles*. This arrangement is

Figure 8-1 High-Speed Synchronous Generator Rotor (Photo courtesy of Canadian General Electric Company).

shown in Fig. 8-1, where the support blocks holding the field winding in position can be seen. Often, when the machine is intended for high-speed applications, the field winding is embedded in slots and the rotor is cylindrical. The model of the cylindrical rotor machine is simpler and will be developed first and applied to generator and motor operation. The model of the salient-pole machine will then be developed separately. The rotor and stator of a typical motor are shown in Fig. 8-2 prior to final assembly. The connections that can be seen at the surface of the rotor form the damping winding, which is similar to the squirrel cage of an induction motor. It is required if a synchronous motor is to be self-starting.

Figure 8-2 Synchronous Motor (Photo courtesy of Canadian General Electric Company).

8.2 FIELD RELATIONSHIPS

Figure 8-3 shows a schematic diagram of a two-phase armature winding on a stator (coils x and y) with a field winding on a rotor (coil f). This arrangement is the simplest form of polyphase synchronous machine. As with the induction machine, there is less algebra involved when demonstrating the basic properties using a two-phase system rather than the more common three-phase system. All coils in the machine are assumed to produce a two-pole sinusoidal field distribution so that we may apply the ideas presented in Sec. 7.2 directly to the armature winding. However, in this case it is convenient to develop the model in terms of generator operation and make the necessary change in polarity for motor operation at the appropriate time.

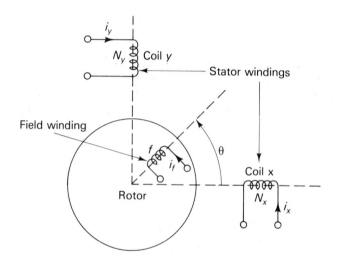

Figure 8-3 Schematic Diagram of Synchronous Generator.

In this case it is the field winding, carrying a constant current, I_f, and being rotated at ω rad/s which produces the magnetic field. This must be a traveling sinusoidal wave, and so Eq. (7-3) may be used directly, the only change required being the subscript to indicate that it is the field winding that is producing this field.

$$\mathcal{F}_f(\theta, t) = \mathcal{F}_{fm} \cos (\theta - \omega t) \tag{8-1}$$

where $\mathcal{F}_{fm}$ is the peak value of the mmf wave. If the air gap is uniform, as in the induction motor, the flux density will have the same form, so that we may write

$$B_f(\theta, t) = B_{fm} \cos (\theta - \omega t) \tag{8-2}$$

The flux linking coil x must then be given by

$$\phi_{xf} = \phi_{xm} \cos \omega t \tag{8-3}$$

and the induced voltage drop is

$$e_x = -N_x \phi_{xm} \sin \omega t$$

For generator operation, where it is usual to assign polarities such that positive current leaves the terminal at which the induced voltage is positive, it is simpler to use the voltage rise rather than the voltage drop. We shall therefore take the source voltage induced in coil x as

$$e_x = N_x \phi_{xm} \sin \omega t \tag{8-4}$$

Note that the angular frequency of the induced voltage is determined directly from the angular velocity of the rotor. We must now investigate the position of the armature mmf waveform relative to the field produced by the field winding. In a synchronous machine, the armature reaction effect is somewhat more complicated than in a dc machine. The current flowing in the armature winding will be sinusoidal, but its timing relative to the voltage in each phase is determined by the power factor of the load. Three principal phase angles will now be considered as examples before attempting to generalize the relative position of the fields (in space) which results from the phase angle (that is, the relative position in time) of the armature current.

1. *Armature current leading voltage by 90°.* This presumes the possibility of a perfectly capacitive load. The current in coil x would be

$$i_x = I_{am} \cos \omega t$$

and the current in coil y must lag this by 90°

$$i_y = I_{am} \sin \omega t$$

Referring to Sec. 7.2, where the two currents were the same as I_x and I_y, the field produced by the armature winding consisting of coils x and y must be the same as given by Eq. (7-3), namely

$$\mathcal{F}_a(\theta, t) = \mathcal{F}_{am} \cos (\theta - \omega t) \tag{8-5}$$

Other than the different magnitudes, the two mmf patterns $\mathcal{F}_f$ and $\mathcal{F}_a$ are exactly aligned, and therefore the resultant mmf waveform may be obtained by adding the magnitudes directly.

2. *Armature current lagging voltage by 90°.* The current in coil x is now the negative of that when the current is purely capacitive.

$$i_x = -I_{am} \cos \omega t$$

Since i_y is also the negative of that which has just been considered, the total mmf produced by the armature winding is the negative of that given by Eq. (8-5). That is,

$$\mathcal{F}_a(\theta, t) = -\mathcal{F}_{am} \cos (\theta - \omega t) \tag{8-6}$$

A purely inductive load produces an armature mmf that is directly opposed to that produced by the field winding, so that the resultant mmf waveform is obtained by subtracting the magnitude.

    3. *Armature current in phase with the voltage.* The phase currents in the armature winding are now

$$i_x = I_{am} \sin \omega t$$

$$i_y = -I_{am} \cos \omega t$$

This set of currents was not considered in Chapter 7, but the total mmf produced by them is obtained simply from

$$
\begin{aligned}
\mathscr{F}_a(\theta, t) &= N_x i_x \cos \theta + N_y i_y \sin \theta \\
&= N_a I_{am} (\cos \theta \sin \omega t - \sin \theta \cos \omega t) \\
&= \mathscr{F}_{am} \sin (\omega t - \theta) \\
&= \mathscr{F}_{am} \cos (90° + \theta - \omega t)
\end{aligned}
\tag{8-7}
$$

This is another sinusoidal distribution of field moving in the positive direction with angular velocity ω, but following the main field at an interval corresponding to an angle of 90°. That is, the armature field lags the field produced by the field winding by 90°. Note that the coil y lags coil x in the air gap because the axis of the field winding passes the axis of coil x before it passes that of coil y. The position of the armature field is therefore consistent with that of the armature field of the dc machine, where the armature current is constrained to be in phase with the induced voltage.

    The actual field in the air gap results from the sum of the two mmf distributions, and the voltage induced in the armature must be determined by this resultant field. It would appear that until the position of the resultant field is known, it is not possible to position the armature mmf in the gap and, until the resultant mmf wave has been positioned, it is not possible to position the armature mmf. Fortunately, this is not the case, and a circuit model can be obtained by using the principle of superposition. The main field and the armature mmf waves are considered to produce separate distributions of flux in the air gap. These flux distributions produce separate induced voltages, and the resultant induced voltage is the sum of these two components. Evidently, when the core is saturated, this model must have some errors. Nevertheless, it is used with correction factors applied to the components of the field. To develop this circuit model we shall first examine phasor diagrams for the three main situations we have just examined. In these diagrams we shall combine the space relationship of the mmf waves and the time relationship of the current waves.

    Although most readers will have met phasor diagrams only in the context of currents and voltages that are sinusoidal functions of time, the method is equally applicable to any variable that is a sinusoidal function of position such as mmf and flux density in the air gap of a machine. In the same manner that a sinusoidal current

$$i(t) = \cos(\omega t + \alpha) = Re[e^{j(\omega t + \alpha)}]$$

may be represented by a line on a complex plane at the angle α relative to the main axis or reference phasor, an mmf distribution

$$\mathcal{F}(\theta) = \cos(\theta + \alpha) = Re[e^{j(\theta + \alpha)}]$$

may also be represented by a line at angle α.

When this is applied to Eqs. (8-1) and (8-5), the phasor representing the mmf produced by the field winding is shown on the real axis, and that for the armature mmf must also be shown in the positive direction of the real axis. From Eq. (8-3) we can see that the flux linking the reference phase of the winding may also be represented on a (time) phasor diagram by a line on the real axis, and from Eq. (8-4) the induced voltage must be shown on the negative imaginary axis. These are shown in Fig. 8-4, where the time and space phasors are shown separately at this stage. Note that since the current and voltage for phase x are taken as representative of the

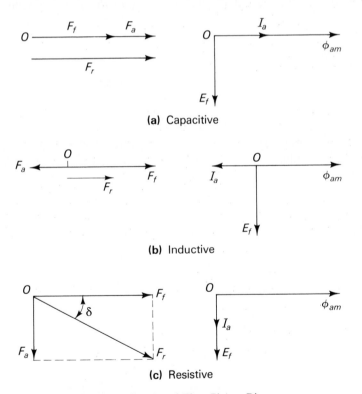

Figure 8-4   Space and Time Phasor Diagrams.

complete winding, it is the flux linking phase $x(\phi_{xm})$ which is taken as that linking the armature winding (ϕ_{am}). With this convention, the (space) mmf phasor produced by the armature current is aligned with its (time) phasor, and that the voltage induced in the reference phase lags (in time) by 90° the flux (time) phasor which is aligned with the (space) mmf phasor that produces this component of flux. Figure 8-4 also shows the corresponding phasor diagrams for purely inductive and purely resistive loading. Unless otherwise specified, the symbols $\mathscr{F}_a$ and $\mathscr{F}_f$ will now be the phasor representation of the armature and field mmf distributions.

The above discussion has been limited to a two-pole machine for clarity. It can be modified for the more general case where there are several pairs of poles. The synchronous speed is now numerically the same as the mechanical speed because of the synchronous nature of these machines. The angular frequency of the currents and voltages is related to the synchronous speed in the same way as was the case of the induction machine. Equation (7-4) is therefore applicable to the synchronous machine; that is, the synchronous speed equals the frequency of the source divided by the pole pairs.

8.3 SYNCHRONOUS GENERATORS

The time and space relationships that were obtained in the previous section can now be combined for any phase angle and arranged in a form that will be interpreted in terms of a circuit model. Figure 8-5 shows a phasor diagram for the case when the armature current lags the induced voltage.

We may now extend the process to consider that each of the mmf distributions produces a separate field distribution with the resultant given by

$$\hat{\mathscr{F}}_r = \hat{\mathscr{F}}_f + \hat{\mathscr{F}}_a \tag{8-8}$$

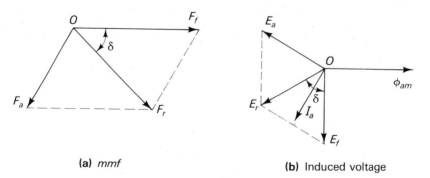

(a) *mmf*   (b) Induced voltage

Figure 8-5 Phasor Diagrams.

and thus two corresponding components of induced voltage with the resultant given by

$$\hat{E}_r = \hat{E}_f + \hat{E}_a \tag{8-9}$$

$\hat{E}_f$ is the *excitation voltage* (sometimes called the *field voltage,* but care is required not to confuse it with the voltage across the field winding) and is the component of induced voltage considered to be produced by the field current. $\hat{E}_a$ is the component of voltage due to the armature current.

Examination of the relative position of the current and voltage phasors in Fig. 8-5 shows that Eq. (8-10) correctly describes this relationship and Fig. 8-6 shows a circuit having this equilibrium equation. It is therefore the circuit model of the ideal synchronous generator. Equation (8-9) is usually expressed as

$$\hat{E}_r = \hat{E}_f - jX_a\hat{I}_a \tag{8-10}$$

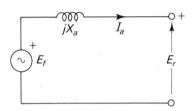

Figure 8-6   Circuit Model of Ideal
Synchronous Generator.

where X_a is the armature reactance, modeling the effect of armature reaction.

The other effects to be included in the circuit model are flux leakage and resistance of the armature winding. In large synchronous machines there is significant skin effect in the armature winding conductors, and the effective or ac resistance must be used. Figure 8-7 shows the complete circuit model. The two reactances are combined and called the *synchronous reactance*

$$X_s = X_a + X_l \tag{8-11}$$

where X_a is the reactance corresponding to the mutual flux due to I_a, and X_l is the leakage reactance. The synchronous reactance derives from the

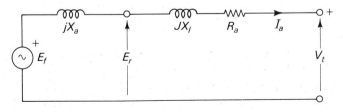

Figure 8-7   Circuit Model of Synchronous Generator.

self-inductance of the armature winding, and therefore this parameter has a profound effect on the characteristics of the synchronous machine. Typical armature resistance values vary from 1% to 5%, leakage reactance values are from 2% to 10%, but synchronous reactance values are from 80% to 150%. With these additions, Eq. (8-10) becomes

$$\hat{E}_f = \hat{V}_t + \hat{I}_a(R_a + jX_s) \tag{8-12}$$

The discussion so far has assumed that the magnetic core is unsaturated. In practice this is not likely to be the case unless the power factor of the armature current is such that it produces a significant demagnetizing effect. Nevertheless, it is common to use this model and, when necessary, adjustments are made in the summation of the mmf waves. These are beyond the scope of this text. However, it is possible to make some allowance for saturation of the core without this complication, recognizing that once again approximations are involved, thus producing approximate predictions of performance.

As with the dc machine, it is best to start by considering the no-load magnetization characteristic. Figure 8-8 shows a typical curve on which the linear portion has been extended. Since the slope is determined primarily by the reluctance of the air gap, it is usual to refer to it as the air-gap line. From a practical consideration, this characteristic may be obtained by running the synchronous machine at its normal operating speed and measuring the open circuit voltage as the field current is varied. Since the armature current is zero, it is therefore a plot of E_f as a function of field current.

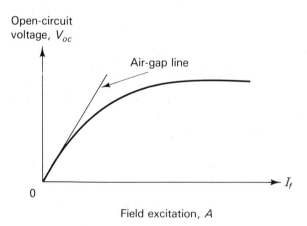

Figure 8-8 Open Circuit Characteristic.

It is also possible to perform a similar test when the armature winding is short-circuited. In this case, due to the large self-inductance of the armature winding, the current lags the induced voltage by almost 90°, and the armature mmf therefore has a demagnetizing effect. Figure 8-9 shows

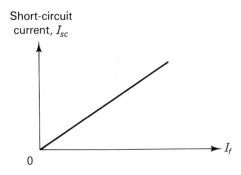

Figure 8-9   Short Circuit Characteristic.   Field excitation, *A*

a typical short circuit characteristic where the core remains unsaturated as a result of this demagnetizing effect.

Also, because of the fact that the synchronous reactance is much greater than the armature resistance, the ratio of open circuit voltage to short circuit current at a particular value of field current may be taken as the synchronous reactance. That is,

$$X_s = \frac{V_{oc}}{I_{sc}} \qquad (8\text{-}13)$$

Due to the saturation of the open circuit characteristic but not of the short circuit characteristic, it is possible to plot the synchronous reactance as a function of field current. This is shown in Fig. 8-10, where at low values of field current it is constant (the unsaturated synchronous reactance) and as the field current is increased its value decreases. When a value is quoted from this region, it is usual to call it the saturated synchronous reactance, but it is important to recognize that such a value is valid only for one

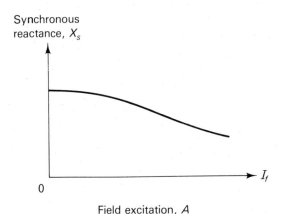

Field excitation, *A*

Figure 8-10   Variation of Synchronous Reactance.

particular value of field current. As a result, the saturated synchronous reactance is of limited value.

Having established the circuit model of a synchronous generator, we may now consider the problem of predicting voltage regulation. The essential part of this problem is to obtain a value for the open circuit voltage corresponding to any given load condition. At first sight it would appear that the excitation voltage in Fig. 8-7 may be used as the open circuit voltage given by

$$\hat{E}_f = \hat{V}_t + \hat{I}_a(R_a + jX_s) \tag{8-14}$$

$$= V_t + I_a(\cos\beta + j\sin\beta)(R_a + jX_s)$$

where the terminal voltage, $\hat{V}_t$, has been taken as the reference phasor and the power factor is $\cos\beta$ (leading). It is true that the magnitude of this voltage would appear to be the open circuit voltage, but such is not the case. If an attempt is made to locate this value on the open circuit characteristic, it will be found only at an excessively high value of field current, which clearly is in error.

To get the open circuit voltage we must first obtain the field current, and this must be done in a manner consistent with the unsaturated synchronous reactance used in the model. The unsaturated synchronous reactance is based on an open circuit characteristic that consists of only the air-gap line in Fig. 8-11, and therefore the value of E_f must be interpreted as a point a on this line rather than on the actual open circuit characteristic. If we use the air-gap line to determine the field current required and then use this value to obtain the open circuit voltage from the actual open circuit characteristic (point b), we obtain a value which, although not accurate, is quite close to the actual value.

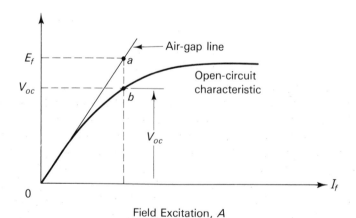

Field Excitation, A

Figure 8-11 Determination of Voltage Regulation.

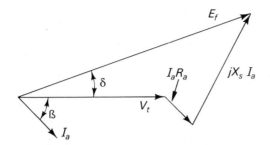

Figure 8-12   Generator Phasor Diagram with Lagging Power Factor.

To complete the discussion of generator operation, phasor diagrams have been shown in Fig. 8-12 for a load that has a lagging power factor and in Fig. 8-13 for a load that has a leading power factor. The main points to observe concern the magnitude of the voltage, E_f, and the sign of the angle δ between $\hat{E}_f$ and $\hat{V}_t$, which is often called the *torque angle*. With a lagging power factor, the excitation voltage generally has a magnitude greater than that of the terminal voltage. With a leading power factor, this magnitude is generally less than that of the terminal voltage. In both situations the excitation voltage leads the terminal voltage and the angle δ is positive. A capacitive load can therefore cause the terminal voltage of a synchronous generator to rise significantly above its rated value, and care is required in its operation if this is likely to happen.

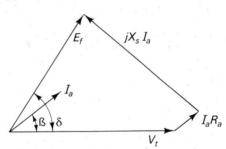

Figure 8-13   Generator Phasor Diagram with Leading Power Factor.

Example 8-1

A three-phase, 11-kV, 60-Hz, 20-MVA, wye-connected, cylindrical rotor synchronous generator has a synchronous reactance of 5 Ω per phase and armature resistance of 0.5 Ω per phase. Determine the excitation voltage when it delivers rated load at 11 kV and 0.9 lagging power factor.

Solution.   For this problem we can refer to Fig. 8-12 and determine all values quite simply. The terminal voltage is chosen as reference, taking care to use the phase value.

$$V_t = \frac{11\ 000}{\sqrt{3}} = 6350.9\ \text{V}$$

The armature current is obtained as usual by dividing the apparent power by the line voltage and by $\sqrt{3}$ so that

$$I_a = \frac{20 \times 10^6}{\sqrt{3} \times 11 \times 10^3} = 1049.7 \text{ A}$$

The phasor representation of the armature current is therefore

$$\hat{I}_a = 1049.7(0.9 - j0.436) = 944.7 - j457.6$$

The voltage equation for generator operation is

$$\begin{aligned}
\hat{E}_f &= \hat{V}_t + \hat{I}_a(R_a + jX_s) \\
&= 6350.9 + (944.7 - j457.6)(0.5 + j5.0) \\
&= 9111.3 + j4494.7 \\
&= 10\ 159.6 \underline{/26.3} \text{ V}
\end{aligned}$$

The excitation voltage is therefore 10 160 V per phase or 17 597 V (line). The angle of 26.3° is also of importance, but detailed discussion has been placed in Sec. 8.5.

If this value of excitation voltage were to be used directly to estimate the voltage regulation, it would be calculated as 60%, but this would be seriously in error. As noted above, the quickest way to get an approximate value is to use the magnetization characteristic to determine the field current at which the voltage on the air-gap line is 17 597 V (i.e., point a in Fig. 8-11). The open circuit voltage is read from the actual open circuit characteristic at this value of field current (i.e., point b in Fig. 8-11).

8.4 SYNCHRONOUS MOTORS

Once again, it is convenient to change the polarity of the current in the circuit model to reflect the fact that the flow of energy is into the armature winding. Figure 8-14 shows the standard model of a synchronous motor in which the air gap is uniform. The voltage equation is therefore

$$\hat{E}_f = \hat{V}_t - \hat{I}_a(R_a + jX_s) \tag{8-15}$$

It is important to note that as long as a synchronous motor operates in its synchronous mode, the steady-state speed will be constant provided the frequency of the source remains constant. Since it is essentially the same machine as a synchronous generator, but with the power flow reversed, we should expect that any change in field current will *not* have any effect on the speed. It will certainly change the value of the excitation voltage

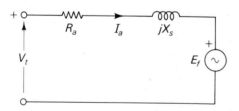

Figure 8-14 Circuit Model of Synchronous Motor.

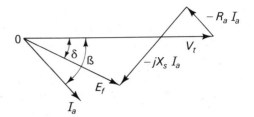

Figure 8-15 Phasor Diagram of Motor
with Lagging Power Factor.

and, as will be shown later, it is the power factor that changes as a result. Before considering this characteristic of the synchronous motor in more detail, the phasor diagrams corresponding to Figs. 8-12 and 8-13 will be examined.

Figure 8-15 shows the phasor diagram of a motor when set to operate with a lagging power factor. The length of the armature resistance voltage drop phasor has been shown larger than it should be in order to make it visible. The phasor diagram for capacitive operation is shown in Fig. 8-16, and it is evident that many combinations of load and leading power factor can result in the magnitude of the excitation voltage being greater than that of the terminal voltage.

If we now examine all four phasor diagrams we must conclude that the only difference between generator and motor operation on these diagrams is that the excitation voltage leads the terminal voltage when generating and lags when motoring. This is consistent with the single-phase synchronous motor, which was considered in Example 4-3, where the angle δ is closely related to the phase angle between these two voltages (it corresponds exactly to the angle between $\hat{E}_f$ and $\hat{E}_r$). From the expressions obtained in that example, it is evident that if the angle δ becomes negative, the torque will reverse although the speed remains the same. In the case of the polyphase synchronous machine, it is useful to think in terms of the machine responding to a change in load by means of a transient change in speed until the rotor is positioned such that the angle δ has acquired the value necessary to

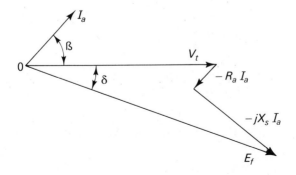

Figure 8-16 Phasor Diagram of Motor with Leading Power Factor.

maintain synchronous operation. This point can be best appreciated by observing the shaft of a synchronous machine under the light of a stroboscope which flashes at the exact synchronous speed and at the same point in the period of the terminal voltage. If the machine is motoring, an increase in load will cause the shaft to appear to rotate in the reverse direction until the new steady-state equilibrium position has been reached.

8.5 POWER-ANGLE RELATIONSHIPS

The power flow in a synchronous machine can be obtained using the normal expressions for the sinusoidal steady state, but it is much more convenient and informative to develop an expression for power in terms of the terminal voltage, the excitation voltage, and the phase angle between them. Figure 8-17a shows the relevant parts of the circuit model with polarity corresponding to generator operation.

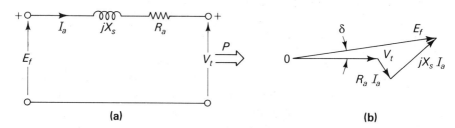

Figure 8-17   Power in a Synchronous Generator.

The solution is based on the geometry of the corresponding phasor diagram, Fig. 8-17b. The excitation voltage is

$$\hat{E}_f = E_f(\cos \delta + j \sin \delta)$$

and the current is

$$
\begin{aligned}
\hat{I}_a &= \frac{\hat{E}_f - \hat{V}_t}{R_a + jX_s} \\
&= \left[\frac{E_f \cos \delta - V_t + jE_f \sin \delta}{R_a + jX_s} \right] \\
&= \left[\frac{[(E_f \cos \delta - V_t) + jE_f \sin \delta](R_a - jX_s)}{R_a^2 + X_s^2} \right] \\
&= \frac{1}{Z^2} [R_a(E_f \cos \delta - V_t) + X_s E_f \sin \delta + jR_a E_f \sin \delta - jX_s(E_f \cos \delta - V_t)]
\end{aligned}
$$

where

$$Z^2 = R_a^2 + X_s^2$$

We now use Eq. (2-5) to obtain the complex voltamperes, noting that $\hat{V}_t$ is the reference phasor.

$$\hat{S} = \hat{V}_t \hat{I}_a^* = P + jQ$$

from which the power and reactive power transmitted per phase are

$$P = \frac{1}{Z^2}[R_a(V_t E_f \cos \delta - V_t^2) + X_s V_t E_f \sin \delta] \tag{8-16}$$

and

$$Q = \frac{1}{Z^2}[-X_s(V_t E_f \cos \delta - V_t^2) + R_a V_t E_f \sin \delta] \tag{8-17}$$

To find the gap power the $I_a^2 R_a$ losses must be added to the value found from Eq. (8-16) in the case of generator operation and subtracted in the case of a motor. Note that for motor operation the angle δ is negative. For some purposes, the armature resistance may be neglected; these two expressions simplify to

$$P = \frac{V_t E_f}{X_s} \sin \delta \tag{8-18}$$

and

$$Q = \frac{V_t^2 - V_t E_f \cos \delta}{X_s} \tag{8-19}$$

These simplified expressions will now be used to explore the characteristics of synchronous machines in more detail.

For motoring applications, the expression for power given by Eq. (8-18) is used to obtain the developed torque. This corresponds to the gap power of the induction machine and, when divided by the synchronous speed (in radians per second), the result is an expression for the developed torque.

$$T_d = \frac{V_t E_f}{\omega_s X_s} \sin \delta \tag{8-20}$$

The maximum developed torque is called the *pull-out torque,* and from Eq. (8-20) it can be seen that it is proportional to the terminal voltage and to the excitation voltage. It is therefore advantageous to operate a synchronous motor with a leading power factor, and there are numerous applications where the installation is used also to improve the power factor of an industrial plant because of the leading power factor that is associated with high-excitation voltage.

8.6 EFFECT OF CHANGE IN EXCITATION

Perhaps the most common method of presenting this characteristic of both motor and generator is by means of *V curves.* These are simply plots of the magnitude of the armature current as the field current is varied. To

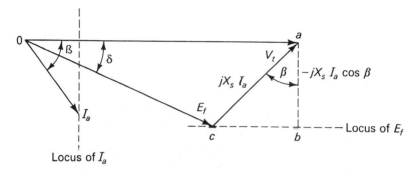

Figure 8-18 Locus of E_f with Varying Excitation.

derive the basic form of these curves, let us consider a synchronous motor with negligible armature resistance. Its voltage equation is thus

$$\hat{E}_f = \hat{V}_t - jX_s\hat{I}_a$$

and Fig. 8-18 shows the locus of the $\hat{E}_f$ phasor as the field current is changed. This locus is explained by first considering the locus of the armature current.

As long as the mechanical system being driven remains unchanged, the torque demanded of the motor will be constant. Since the losses are considered negligible, this means that the in-phase component of $\hat{I}_a$ must also be constant. The locus of the $\hat{I}_a$ phasor must therefore be a line parallel to the imaginary axis. In graphical terms, we subtract the voltage drop across the synchronous reactance by drawing a line from the end of the $\hat{V}_t$ phasor, recognizing that the $I_a \cos \beta$ component of this voltage must also be constant. This locus must therefore be a line parallel to the real axis. In addition, it may be noted by referring to Eq. (8-20) that as the

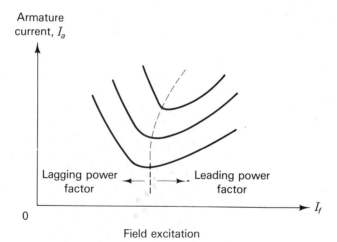

Figure 8-19 Synchronous Motor V-Curves.

excitation is increased, E_f increases and the torque angle must decrease since T_d, V_t, and X_s are constant. This is evident in Fig. 8-18. The magnitude of the corresponding armature current can be determined from the I_aX_s voltage and plotted as a V curve in Fig. 8-19.

If the power demanded by the mechanical system is increased, the locus will move farther away from the axis. Examination of the locus shows that at low values of excitation the armature current lags the terminal voltage, but as the excitation is increased the motor takes on the characteristic of a capacitive circuit. This property of the synchronous motor makes it attractive for driving loads that are connected for long periods. The power factor of an industrial plant can be improved, while at the same time useful mechanical output is obtained. A common application is that of driving a compressor such as the installation shown in Fig. 8-20.

In the case of generator operation the development is the same except that the excitation voltage leads the terminal voltage. Perhaps the most important conclusion regarding the operation of a synchronous generator forming part of a large power system is that its controls do not perform the functions that they would if the generator were completely on its own. The voltage of an isolated generator can be controlled by changing the magnitude of the field current, and the rheostat or control circuit that does this is therefore called a *voltage regulator*. When the same generator and voltage regulator are connected to a system having other synchronous generators, it is only a slight exaggeration to say that the voltage regulator has absolutely no control over the terminal voltage, but it does control the power factor. Similarly, the speed regulator of an isolated synchronous

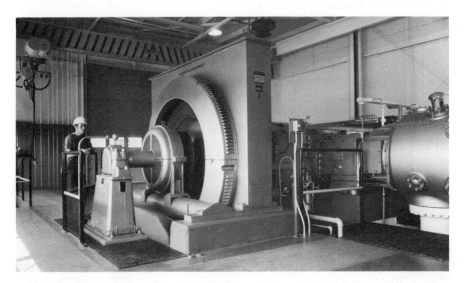

Figure 8-20 Synchronous Motor Driving a Reciprocating Compressor (Photo courtesy of Canadian General Electric Co.).

generator will control its speed, but when the generator forms part of a larger system it controls only the power.

Example 8-2

A three-phase, 460-V, 60-Hz, ten-pole, wye-connected, cylindrical rotor synchronous motor has a synchronous reactance of 2.3 Ω per phase and negligible armature resistance.

(a) Determine the excitation voltage, E_f, if the input power is 50 kW at 0.8 leading power factor when connected to a 460-V source.

(b) Assuming that the excitation voltage is proportional to field current, determine the effect on the armature current of a 10% decrease in field current.

Solution

(a) The first part is virtually the same as the solution for Example 8-1, the only difference being that this is a motor rather than a generator. The terminal voltage is

$$V_t = \frac{460}{\sqrt{3}} = 265.6 \text{ V}$$

The armature current is

$$I_a = \frac{50 \times 10^3}{\sqrt{3} \times 460 \times 0.8} = 78.44 \text{ A}$$

or

$$\hat{I}_a = 78.44(0.8 + j0.6) = 78.44 \underline{/36.87} = 62.75 + j47.06 \text{ A}$$

The excitation voltage is thus

$$\hat{E}_f = 265.6 - (62.75 + j47.06)(j2.3) = 265.6 + 108.2 - j144.3$$
$$= 373.8 - j144.3 = 400.7 \underline{/-21.1} \text{ V (phase)}$$

(b) The key to the solution is the fact that when the losses are neglected, the power and developed torque remain the same. From Eq. (8-20) we can see that any decrease in the excitation must cause an increase in the torque angle (sin δ) since all other terms are constant. In terms of the locus shown in Fig. 8-18, the imaginary component of $\hat{E}_f$ remains $-j144.3$ V, while its magnitude is reduced to 90% of 400.7, namely 360.63 V. Thus its real component is

$$E_{fr} = \sqrt{(360.63^2 - 144.3^2)} = 330.5 \text{ V}$$

and the torque angle is $\tan^{-1}(144.3/330.5) = 23.59°$. This may be compared with the value obtained by using Eq. (8-18) directly. That is,

$$\sin \delta = \frac{50 \times 10^3 \times 2.3}{3 \times 265.6 \times 360.63} = 0.4002$$

so that the torque angle is 23.59°.

The armature current is obtained by solving the voltage equation, noting that phasor notation is essential.

$$\hat{E}_f = \hat{V}_t - jX_s\hat{I}_a$$

Thus

$$jX_s\hat{I}_a = 265.6 - 330.5 + j144.3 = -64.9 + j144.3 = 158.2 \underline{/114.2}$$

and

$$\hat{I}_a = \frac{158.2}{2.3} \underline{/114.2 - 90}$$

$$= 68.78 \underline{/24.2}$$

Since the motor was overexcited to start with, the result of the small decrease in excitation is a reduction in the magnitude of the armature current from 78.44 A to 68.78 A, a reduction in the phase angle between $\hat{V}_t$ and $\hat{I}_a$ from 36.87° to 24.2°, and an increase in the torque angle from 21.2° to 23.59°.

8.7 SALIENT POLE MACHINES

Although most large high-speed synchronous generators are made with cylindrical rotors which provide an air gap that is substantially uniform, it is common to employ salient poles in low-speed synchronous motors and generators. The field coils are mounted on pole pieces which cover approximately 70% of the periphery. This may be seen in Fig. 8-2. The armature windings are placed in slots which still leave a surface that is virtually cylindrical. As a result, the model that has provided the basis for our discussion so far is not adequate for the salient-pole machine. A synchronous motor for use in a steel mill is shown in Fig. 8-21, where the slip

Figure 8-21 Synchronous Motor for Use in a Steel Mill (Photo courtesy of Canadian General Electric Co.).

Figure 8-22 Synchronous Motor with Brushless Exciter (Photo courtesy of Canadian General Electric Co.).

rings may readily be seen. There are arrangements possible whereby the slip rings are not required; Fig. 8-22 shows an example of a brushless synchronous motor.

To understand the basis of the model of the salient-pole machine, it is necessary to return to Sec. 8.2 in which the field relations were described. Figure 8-23 shows the new arrangement; we must now consider the position of the armature field relative to the poles of the main field. We have seen that the relative position of the armature field is determined by the (time) phase angle of the armature current. There are conditions where the two

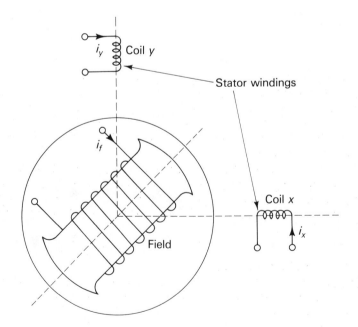

Figure 8-23   Schematic Diagram of Salient Pole Machine.

fields are aligned and therefore the flux produced by the armature mmf will be at its greatest. There is also the possibility that the armature field is displaced from the main axis of magnetization of the field poles by 90°, and this will produce minimum flux per ampere.

We now define two armature inductances corresponding to these situations. The principal axis of magnetization is called the *direct axis* (d-axis) and the complementary axis is called the *quadrature axis* (q-axis). The reactance X_{ad} is the armature reactance when the armature field is aligned with the d-axis, and the reactance X_{aq} is the armature reactance when the armature field is aligned with the q-axis. The armature current, $\hat{I}_a$, is now considered to be resolved into two components: one ($\hat{I}_{ad}$ or just $\hat{I}_d$) which produces a component of armature field on the d-axis and the other ($\hat{I}_{aq}$ or just $\hat{I}_q$) which produces a component of armature field on the q-axis. The voltage induced in the armature winding is now considered to have three components coming from the three components of field: $\hat{E}_f$ due to the field current I_f, $\hat{E}_{ad}$ due to $\hat{I}_{ad}$, and $\hat{E}_{aq}$ due to $\hat{I}_{aq}$. The phasor diagram for this is shown in Fig. 8-24, where the excitation voltage is taken as reference phasor. The resultant induced voltage is the sum of the three components

$$\hat{E}_r = \hat{E}_f + \hat{E}_{ad} + \hat{E}_{aq} \tag{8-21}$$

$$= \hat{E}_f - jX_{ad}\hat{I}_d - jX_{aq}\hat{I}_q$$

The armature current may also be considered as the sum of two components

$$\hat{I}_a = \hat{I}_{aq} + \hat{I}_{ad} \tag{8-22}$$

$$= \hat{I}_q + j\hat{I}_d$$

but this is of limited value in this form, and it is better to examine the particular phasor diagram when combining the components to get the armature current.

Equation (8-21) shows the two components of voltage adding, a property associated with a series circuit. In contrast, Eq. (8-22) shows the two components of current adding, normally modeled by means of a parallel circuit. As a result, these two constraints are incompatible with any simple

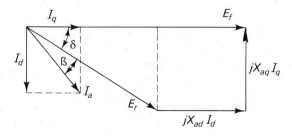

Figure 8-24   Phasor Diagram of Ideal Salient Pole Generator.

circuit; therefore there is no simple-circuit model of the salient-pole synchronous machine. The circuit model of the cylindrical rotor machine may
be used as an aid in formulating the equations that follow, but unfortunately
it is of no further use when dealing with salient-pole machines.

The leakage flux and armature resistance are now added to the above
equations to complete the model. Since both components of current are
affected by the leakage reactance, we now define the normal reactances as
follows.

$$X_d = X_{ad} + X_l \text{ is the } direct \text{ axis reactance}$$

and

$$X_q = X_{aq} + X_l \text{ is the } quadrature \text{ axis reactance}$$

With these terms now replacing X_{ad} and X_{aq} in Eq. (8-21), the voltage
equation for the salient-pole generator becomes

$$\hat{V}_t = \hat{E}_f - jX_d\hat{I}_d - jX_q\hat{I}_q - R_a\hat{I}_a \tag{8-23}$$

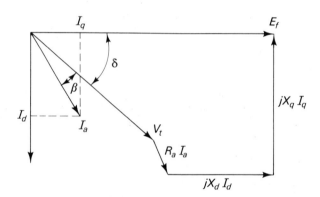

Figure 8-25   Phasor Diagram of Salient Pole Generator.

Figure 8-25 is a phasor representation of this equation. The solution
of this equation poses a severe problem. It appears that until we find the
position of the $\hat{E}_f$ phasor relative to the terminal voltage we cannot split
the armature current into its two components, but without these two components we cannot place the $\hat{E}_f$ phasor. There are several solutions found
in different texts. They are variations on what is essentially a graphical
solution. Consider the more detailed version of the phasor diagram shown
in Fig. 8-26, where a triangle $O'a''b''$ has been formed by drawing $O'a''$
perpendicular to the armature current phasor. The triangles Oab, $O'a'b'$
and $O'a''b''$ are all similar. The line $a'b'$ represents $X_q I_q$ and therefore

$$O'a' = \frac{X_q I_q}{\cos(\beta + \delta)}$$

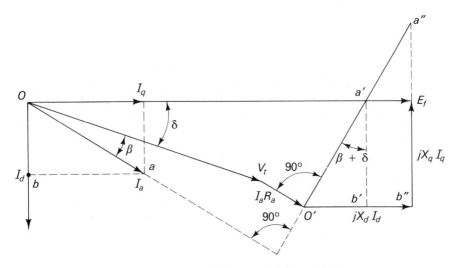

Figure 8-26   Detailed Phasor Diagram of Salient Pole Generator.

Examination of the triangle Oab shows that $I_a = I_q/\cos(\beta + \delta)$, and so

$$O'a' = X_q I_a$$

This is how the $\hat{E}_f$ phasor is placed relative to the terminal voltage phasor. Having now placed $\hat{E}_f$, we can resolve $\hat{I}_a$ into its two components and use Eq. (8-21) directly. In this form, the solution does not require any scale drawings, but it is important to refer frequently to the phasor diagram. That is, although the solution is analytical in nature, it is based on the geometry of the phasor diagram. However, it is possible to continue graphically by showing that $O'a''$ equals $X_d I_a$, and by drawing a line perpendicular to $\hat{E}_f$, we fix its magnitude as well as its direction.

8.7.1 Torque Developed by Salient-Pole Machines

The expression corresponding to Eq. (8-20) is obtained by interpretation of the phasor diagram shown in Fig. 8-25. The terminal voltage, $\hat{V}_t$, is resolved along the d- and q-axes so that each of these components is in phase with $\hat{I}_d$ and $\hat{I}_q$, the two components of the armature current.

$$P = I_d V_t \sin \delta + I_q V_t \cos \delta \qquad (8\text{-}24)$$

Examination of the phasor diagram gives the following expressions for I_d and I_q when R_a is negligible.

$$\sin \delta = \frac{I_q X_q}{V_t}$$

from which

$$I_q = \frac{V_t}{X_q} \sin \delta \qquad (8\text{-}25)$$

and

$$\cos \delta = \frac{E_f - I_d X_d}{V_t}$$

from which

$$I_d = \frac{E_f - V_t \cos \delta}{X_d} \tag{8-26}$$

Substituting Eqs. (8-25) and (8-26) into Eq. (8-24) gives the following expression for the developed power.

$$P = \left[\frac{E_f - V_t \cos \delta}{X_d}\right] V_t \sin \delta + \frac{V_t}{X_q}[\sin \delta \; V_t \cos \delta]$$

$$= \frac{E_f V_t}{X_d} \sin \delta + V_t^2\left(\frac{1}{X_q} \sin \delta \cos \delta - \frac{1}{X_d} \sin \delta \cos \delta\right)$$

$$= \frac{E_f V_t}{X_d} \sin \delta + \frac{V_t^2}{2}\left(\frac{1}{X_q} - \frac{1}{X_d}\right) \sin 2\delta \tag{8-27}$$

Figure 8-27 shows a plot of Eq. (8-27) where the increased initial slope is evident. The position of the second harmonic term is always such that the salient-pole machine is "stiffer" than the corresponding cylindrical rotor machine. That is, the transient changes in speed as the load changes are slightly less. The developed torque is obtained as before, by dividing the developed power by the synchronous speed (in radians per second).

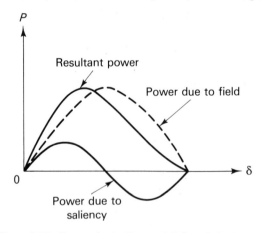

Figure 8-27   Power Angle Curve of Salient Pole Generator.

Example 8-3

A three-phase, 2300-V, 60-Hz, 16-pole, wye-connected, salient-pole synchronous motor has a direct axis reactance of 3.5 Ω per phase and a quadrature axis reactance of 2.5 Ω per phase. The armature resistance is to be neglected. If the excitation is such that when delivering 2000 kW the power factor is 0.9 (lagging), determine the excitation voltage and torque angle.

Solution.   The first part of the solution is to obtain the value of the armature current. This is

$$I_a = \frac{2000 \times 10^3}{\sqrt{3} \times 2300 \times 0.9} = 558.83 \text{ A}$$

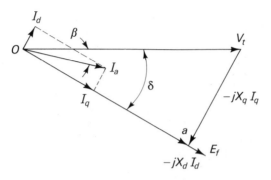

Figure 8-28   Phasor Diagram, Example 8-3.

The phasor diagram of Fig. 8-28 is now constructed using the terminal voltage as reference phasor. First, the armature current whose magnitude has just been obtained must be expressed as a phasor by noting that the current lags the voltage by 25.8° when the power factor is 0.9 (lagging).

$$\hat{I}_a = 558.83 \underline{/-25.8} = 502.05 - j243.15 \text{ A}$$

The direction of the excitation voltage phasor can now be fixed by subtracting the $\hat{I}_a X_q$ voltage drop from the terminal voltage. That is,

$$Oa = \frac{2300}{\sqrt{3}} - j2.5(502.05 - j243.15)$$

$$= 720 - j1255.1 = 1447 \underline{/-60.2} \text{ V}$$

Thus the torque angle is now known to be $-60.2°$. If we wish to proceed without resort to a graphical solution, we must now resolve the armature current into its two components. When substituting values in the voltage equation, it is vital to remember that all currents must be represented by their phasors. This is why the angles for $\hat{I}_d$ and $\hat{I}_q$ are shown in addition to their magnitudes. From the phasor diagram, the phase angle between $\hat{E}_f$ and $\hat{I}_a$ is found to be $60.2 - 25.8 = 34.4°$. The current $\hat{I}_q$ is aligned with $\hat{E}_f$, and it is therefore given by

$$\hat{I}_q = I_a \cos 34.4° \underline{/-60.2} = 460.27 \underline{/-60.2} = 228.74 - j399.41 \text{ A}$$

The current $\hat{I}_d$ evidently leads $\hat{I}_q$ in this case, so that it is given by

$$\hat{I}_d = I_a \sin 34.4° \underline{/90-60.2} = 315.16 \underline{/29.8} = 273.48 + j156.62 \text{ A}$$

Having determined the components of the armature current, we may now substitute their values in the voltage equation to get the excitation voltage.

$$\hat{E}_f = \hat{V}_t - jX_d\hat{I}_d - jX_q\hat{I}_q$$
$$= 1327.9 - j3.5(273.48 + j156.82) - j2.5(228.74 - j399.41)$$
$$= 1763 \ \diagup -60.15 \text{ V}$$

When using this approach, it is wise always to determine the angle associated with $\hat{E}_f$ as a check that the angles attributed to $\hat{I}_d$ and $\hat{I}_q$ have been correctly incorporated and used. The small difference in the value of the phase angle is due to the rounding out of the values during each stage of the calculation. The solution is completed by noting that the excitation voltage is 1763 V per phase or 3054 V (line).

PROBLEMS

8-1. A three-phase, 220-V, 60-Hz, 20-kVA, six-pole, wye-connected, cylindrical rotor synchronous generator supplies rated load at 0.9 lagging power factor and 220 V. The synchronous reactance is 4 Ω per phase and the armature resistance is 0.5 Ω per phase. Its open circuit characteristic at synchronous speed is as follows:

I_f	0	1	2	3	4	5	6	8	10
E_f	0	75	140	190	220	240	255	270	290

Calculate
(a) the excitation voltage,
(b) the field current required,
(c) the voltage regulation.

8-2. A three-phase, 550-V, 60-Hz, 50-kVA, 12-pole, wye-connected, cylindrical rotor synchronous generator supplies rated load at 0.8 leading power factor and 550 V. The synchronous reactance is 6 Ω per phase and the armature resistance is 0.7 Ω per phase. Its open circuit characteristic at synchronous speed is as follows.

I_f	0	2	4	6	8	10	12	16	20
E_f	0	187	350	475	550	600	638	675	725

Calculate
(a) the excitation voltage,
(b) the field current required,
(c) the voltage regulation.

8-3. A three-phase, six-pole, wye-connected, cylindrical rotor synchronous motor is connected to a 220-V, 60-Hz source and takes a current of 150 A at 0.8 leading power factor when driving a particular mechanical system. It has the following magnetization characteristic at synchronous speed.

I_f	0	1.5	3.0	6.0	9.0	12.0	15.0	18.0	21.0	24.0
E_f	0	47	95	146	179	201	217	230	239	247

The unsaturated synchronous reactance is 0.6 Ω per phase and the armature resistance is to be neglected. Determine
(a) the excitation voltage,
(b) the developed torque,
(c) the field current,
(d) the adjustment required to change the power factor to 0.8 (lagging).

8-4. A three-phase, eight-pole, wye-connected, cylindrical rotor synchronous motor is connected to a 460-V, 60-Hz source and takes a current of 350 A at 0.8 lagging power factor when driving a particular mechanical system. It has the following magnetization characteristic at synchronous speed.

I_f	0	2.5	5.0	10.0	15.0	20.0	25.0	30.0	35.0	40.0
E_f	0	94	190	292	358	402	434	460	478	494

The unsaturated synchronous reactance is 1.1 Ω per phase and the armature resistance is negligible. Determine
(a) the excitation voltage,
(b) the developed torque,
(c) the field current,
(d) the adjustment required to change the power factor to 0.8 (leading).

8-5. A three-phase, 6600-V, 60-Hz, wye-connected, cylindrical rotor synchronous motor has a synchronous reactance of 95 Ω per phase. It is operating with an input power of 100 kW at 0.8 leading power factor when connected to a 6600-V source. Determine
(a) the excitation voltage, E_f,
(b) the torque angle, δ.

8-6. A 3-ph, 6600-V, 60-Hz, 8-pole, wye-connected, cylindrical rotor synchronous motor has a synchronous reactance of 45 Ω per phase and armature resistance of 4.0 Ω per phase. It is operating with a developed power of 200 kW when connected to a 6600-V source and the excitation adjusted so that the excitation voltage has the same magnitude as the terminal voltage. Determine
(a) the input power,
(b) the input reactive power,
(c) the developed torque.

8-7. Repeat Problem 8-6, neglecting the armature resistance.

8-8. A three-phase, 2300-V, 60-Hz, 2000-hp, 22-pole, wye-connected, cylindrical rotor synchronous motor has a synchronous reactance of 4.25 Ω per phase and negligible losses. Calculate the maximum torque that can be developed when connected to a 2300-V source and the excitation adjusted to result in unity power factor at rated load.

8-9. A three-phase, 6000-V, 60-Hz, 3000-hp, 26-pole, wye-connected, cylindrical rotor synchronous motor has a synchronous reactance of 5.15 Ω per phase and negligible losses. Calculate the maximum torque that can be developed when connected to a 6000-V source and the excitation adjusted to result in a power factor of 0.8 (leading) at rated load.

8-10. A three-phase, 550-V, 60-Hz, 10-pole, wye-connected, cylindrical rotor synchronous motor has a synchronous reactance of 2.9 Ω per phase and negligible armature resistance.

   (a) Determine the excitation voltage, E_f, if the input power is 75 kW at unity power factor when connected to a 550-V source.

   (b) Assuming that the excitation voltage is proportional to field current, determine the effect on the armature current of a 20% increase in field current.

8-11. A three-phase, 460-V, 60-Hz, ten-pole, wye-connected, cylindrical rotor synchronous motor has a synchronous reactance of 2.5 Ω per phase and negligible armature resistance.

   (a) Determine the excitation voltage, E_f, if the input power is 60 kW at unity power factor when connected to a 460-V source.

   (b) Assuming that the excitation voltage is proportional to field current, determine the effect on the armature current of a 25% increase in field current.

8-12. A three-phase, 460-V, 60-Hz, ten-pole, wye-connected cylindrical rotor synchronous motor has a synchronous reactance of 0.8 Ω per phase. The armature resistance may be neglected. It is driving a system with its excitation set to give a power factor of 0.8 (leading) when the power is 200 kW and the terminal voltage is 460 V.

   (a) Determine the armature current and excitation voltage for this operating condition.

   (b) If the load torque is now doubled, determine the armature current, power, and power factor.

8-13. A three-phase, 2300-V, 60-Hz, ten-pole, wye-connected cylindrical rotor synchronous motor has a synchronous reactance of 4.0 Ω per phase. The armature resistance may be neglected. It is connected to a 2300-V source and drives a system with its excitation set to give a power factor of 0.9 (leading) when the power is 500 kW. Determine

   (a) the armature current,

   (b) the excitation voltage for this operating condition.

   (c) If the load torque is increased by 50%, determine the armature current, power, and power factor.

8-14. A three-phase, 220-V, 60-Hz, 20-kW, eight-pole, wye-connected, salient-pole synchronous motor is operating with an input of 20 kW at 0.9 leading power factor and 220 V. The d-axis reactance is 3 Ω per phase, the q-axis reactance is 2 Ω per phase, and the armature resistance is 0.5 Ω per phase. Determine

   (a) the excitation voltage, E_f,

   (b) the torque,

   (c) the torque angle.

8-15. A three-phase, 600-V, 60-Hz, 75-kW, six-pole, wye-connected, salient-pole synchronous motor is operating with an input of 75 kW at 0.8 leading power factor and 600 V. The d-axis reactance is 5 Ω per phase, the q-axis reactance

is 3.5 Ω per phase, and the armature resistance is 0.5 Ω per phase. Determine
(a) the excitation voltage, E_f,
(b) the torque,
(c) the torque angle.

8-16. A three-phase, salient-pole synchronous motor is operated in an overexcited condition (i.e., the field current is such that the power factor is leading). From the phasor diagram, show that the torque angle is given by

$$\tan \delta = \frac{I_a X_q \cos \beta + I_a R_a \sin \beta}{V_t + I_a X_q \sin \beta - I_a R_a \cos \beta}$$

8-17. A three-phase, 2300-V, 60-Hz, 16-pole, wye-connected, salient-pole synchronous motor has $X_d = 6.8$ and $X_q = 4.8$ Ω per phase. The armature resistance may be neglected. The excitation is adjusted so that the power is 700 kW at unity power factor when the terminal voltage is 2300 V. Determine the pull-out torque corresponding to this excitation.

8-18. A three-phase, 11 000-V, 60-Hz, 28-pole, wye-connected, salient-pole synchronous motor has $X_d = 80.7$ and $X_q = 65.3$ Ω per phase. The armature resistance may be neglected. The excitation is adjusted so that the power is 1200 kW at 0.8 leading power factor when the terminal voltage is 11 000 V. Determine the pull-out torque corresponding to this excitation.

9

ELECTROMECHANICAL MODELS

9.1 INTRODUCTION

In the previous chapters we have examined the main features of the most common electric machines in sufficient detail to obtain a circuit model for steady-state operation. It may now be observed that we have not attempted solutions to problems where the steady-state parameters (e.g., load torque in newton meters) of the mechanical system have been given directly. This is simply because the conventional models presented in Chap. 6 and 7 cannot provide a direct solution for the operating speed, and iterative methods must be used. As a result, they are somewhat cumbersome to use when investigating changes in the mechanical system, although such investigations are sometimes required. However, the standard methods have been used in the absence of any alternative approach.

The most common form of modeling used for dynamic studies regards each machine as a dynamic circuit, and transformations are sought that will linearize the resulting equilibrium equations. References 1 and 2 are examples of texts in which this material may be found. A full appreciation of these models requires more knowledge than may reasonably be gathered in a first course in the subject; therefore, they are usually avoided at this stage. Nevertheless, it is not uncommon to meet a problem where some form of dynamic response is required and some form of solution must be obtained.

A convenient solution to both these difficulties may be found by regarding the machine as a linear system and taking the Thevenin model from its mechanical port or shaft. This is simply an extension of the common practice of using analogous circuits. The difference is that in this case we shall, in principle, be working with a mechanical analog of the electrical system. However, as with many mechanical systems, this mechanical analog will be expressed or described in circuit terms.

The form of analog that has been found most useful is that based on through and across variables. When this is done, it is possible to obtain the analogous circuit of a mechanical system by means of a simple examination of the topology of the system, and then using this analogous circuit to obtain the equilibrium equations of the system. There are many cases where this approach gives the equilibrium equations very rapidly and reliably. Details may be found in several texts such as reference 3.

9.2 DC MOTORS

For dynamic studies of a dc motor, Eq. (6-14) is expanded to include the self-inductance (L_a) of the armature winding. Equation (6-16) for the speed voltage is modified, since we must assume that the flux is proportional to the field current if we are to obtain a linear model. With these changes, the model of the dc motor has the following equilibrium equations.

$$v_a = R_a i_a + L_a \frac{di_a}{dt} + M i_f \omega_m \tag{9-1}$$

$$T_d = M i_f i_a \tag{9-2}$$

where M is the speed mutual inductance that replaces the constants K_t and K_v of Chap. 6, i_a is the armature current, i_f is the field current, v_a is the armature terminal voltage, ω_m is the angular velocity of the armature, and T_d is the developed torque. These two equations can be rearranged as an expression for the armature speed, which will be regarded as a function of the developed torque. In doing so, we must consider the most likely constraints to be applied to the electrical side of the motor. Since the objective is to obtain a Thevenin model for investigation of changes in the mechanical system, we may reasonably take the armature voltage and the field current as constant. They are therefore shown with uppercase symbols.

$$\omega_m = \frac{1}{MI_f} \left[V_a - (R_a + L_a p) \frac{T_d}{MI_f} \right] \tag{9-3}$$

where p is the differential operator, d/dt

$$\omega_m = \omega_o - \frac{1}{(MI_f)^2} (R_a + L_a p) T_d \tag{9-4}$$

If we now recall that an ideal damper has the characteristic that the velocity difference between its two terminals is proportional to the torque transmitted, and that an ideal spring has the characteristic that the velocity difference between its two terminals is proportional to the derivative of the torque transmitted, Eq. (9-4) may be rewritten as

$$\omega_m = \omega_o - \frac{1}{B_o} T_d - \frac{1}{K_o} p T_d \tag{9-5}$$

where

$$B_o = \frac{(MI_f)^2}{R_a} \tag{9-6}$$

and

$$K_o = \frac{(MI_f)^2}{L_a} \tag{9-7}$$

Figure 9-1 is a circuit having its equilibrium described by Eq. (9-5) so that it is a circuit model of the motor. As long as the constraints applied to the electrical terminals remain the same, this circuit may be used as a model of the complete motor (other than its inertia and rotational losses, which will be added later) and the analogous circuit representation of the mechanical system may be connected to its terminals to provide an analogous circuit of the complete electromechanical system.

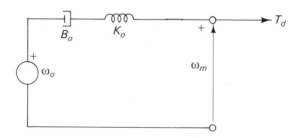

Figure 9-1 Mechanical Circuit Equivalent.

One simple check that may be performed on this circuit is to observe that since it is an exact equivalent of Eqs. (9-1) and (9-2), and since the damper B_o is modeling the effect of the armature resistance, the power dissipated in B_o should be the $I_a^2 R_a$ loss in the armature.

$$P_b = \frac{T_d^2}{B_o} = (MI_f I_a)^2 \times \frac{R_a}{(MI_f)^2} = I_a^2 R_a \tag{9-8}$$

9.2.1 Steady-State Characteristics

For steady-state operation, the circuit model becomes simply that shown in Fig. 9-2, where the damper B_t may include an allowance for the armature reaction effect based on an assumed linear decrease of k (per

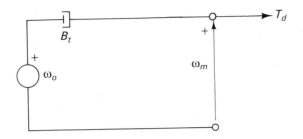

Figure 9-2 Steady-State Circuit Model.

unit) in the flux at a particular value of armature current, I_{ao}. Its value is given by

$$B_t = \frac{B_e}{1 + k'} \tag{9-9}$$

where

$$B_e = \frac{(MI_f)^2}{R_a - kV_a} \tag{9-10}$$

and

$$k' = 2kI_{ao} \tag{9-11}$$

When the armature reaction effect is neglected, B_t is the same as B_o and solutions may be compared directly with those obtained in Chap. 6. There is significant approximation in the derivation of Eqs. (9-9) to (9-11); details may be found in reference 4.

The mechanical load must now be modeled; if its torque is proportional to speed, it is represented by a simple damper with the result that the analogous circuit, shown in Fig. 9-3, is simply a voltage divider. The output speed is obtained simply, noting that a damping parameter, B_L, is analogous to electrical conductance. Hence

$$\omega_m = \left(\frac{B_t}{B_t + B_L}\right)\omega_o \tag{9-12}$$

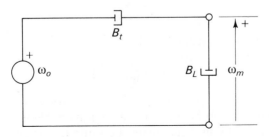

Figure 9-3 Load Torque Proportional to Speed.

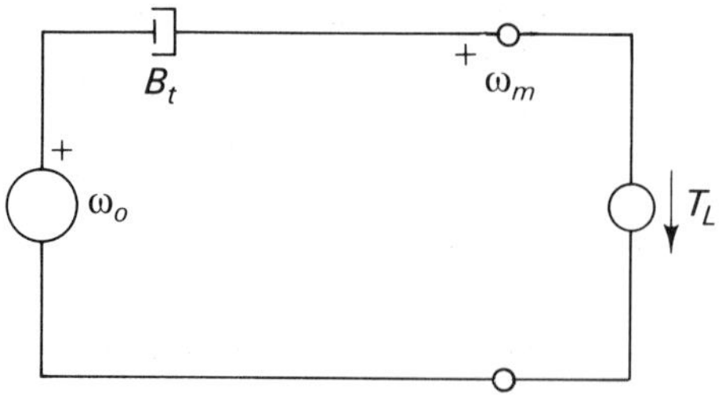

Figure 9-4 Constant Torque Load.

If the mechanical system requires constant load torque, it is modeled by a torque source as shown in Fig. 9-4. Superposition gives the resulting speed as

$$\omega_m = \omega_o - \frac{1}{B_t}T_L \qquad (9\text{-}13)$$

9.2.2 Dynamic Characteristics

A capacitor (analogous to the inertia of the motor) is added to the circuit model of Fig. 9-1 to complete the model when used for dynamic studies. A damper may also be shown to represent the mechanical losses in the motor. Figure 9-5 shows the final model where the analogous spring stiffness, K_t, is related to K_o in the same manner as B_t is related to B_o.

The standard methods of deriving a transfer function may now be used. Perhaps the simplest method to obtain that of Fig. 9-5 is to combine B and J into an equivalent impedance

$$Z(s) = \frac{1}{B + Js} \qquad (9\text{-}14)$$

using the Laplace operator s rather than the differential operator p used at the beginning of the chapter. After some algebraic manipulation the result is

$$\frac{\omega_m}{\omega_o}(s) = \frac{B_t K_t}{B_t J s^2 + (K_t J + B_t B)s + K_t(B_t + B)} \qquad (9\text{-}15)$$

For an armature-controlled motor, we note that

$$\omega_o = \frac{V_a}{MI_f}$$

and the more common transfer function

$$\frac{\omega_m}{V_a}(s) = \frac{1}{MI_f}\left(\frac{\omega_m}{\omega_o}\right) \qquad (9\text{-}16)$$

Referring again to Eqs. (9-1) and (9-2), the products of variables can cause severe analytical problems if at least one variable is not constrained to be constant. In addition, when magnetic saturation is present, the value of M is no longer constant. We are then faced with a set of nonlinear equations for which the Laplace transform is not valid; thus, the ordinary

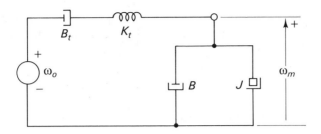

Figure 9-5   Damping and Inertia Load.

methods of linear control theory may not be applied. An example of the analysis of such a system may be found in reference 5.

Example 9-1

A 230-V, 10-hp dc motor has the following parameter values:

$$R_a = 0.3 \ \Omega$$

$$R_f = 188.5 \ \Omega$$

$$M = 1.5 \text{ H}$$

It is operated as a shunt motor with 230 V across both armature and field windings. Neglecting armature reaction, determine the steady-state operating speed when the mechanical load consists of a combination of a constant torque component of 30 N·m and a viscous friction component of 0.25 N·m·s.

Solution.   The first step is to obtain the values of the parameters in Fig. 9-2 when the steady-state field current is $230/188.5 = 1.22$ A. Substituting this value and the speed mutual inductance in Eq. (9-6) gives

$$B_o = 11.163 \text{ N·m·s}$$

and the ideal no-load speed (open circuit speed) is $230/(1.5 \times 1.22) = 125.7$ rad/s or 1200 r/min. The operating speed may be obtained using superposition. The response due to the velocity source is obtained using the ordinary relation for a voltage divider given by Eq. (9-12). That is,

$$\omega_{mo} = \frac{11.136}{11.136 + 0.25} \times 125.7 = 122.94 \text{ rad/s}$$

The change in operating speed due to the constant torque load is

$$\omega_{mT} = -\frac{30}{(11.163 + 0.25)} = -2.63 \text{ rad/s}$$

The operating speed is therefore the algebraic sum of these two components:

$$\omega_m = \omega_{mo} + \omega_{mT} = 122.94 - 2.63 = 120.31 \text{ rad/s or 1148.9 r/min}$$

9.3 INDUCTION MOTORS

From an analytical point of view, the induction motor is more complicated than the dc motor. The approximations made in coming to an electro-mechanical model are more severe and thus limit its application somewhat.

However, the steady-state operating point can be determined with good accuracy, provided it is within the normal working range where the developed torque is proportional to the slip. The electromechanical network model for steady-state operation can be developed on the basis of Chap. 7, but that for dynamic studies requires a knowledge of dynamic circuit analysis.

9.3.1 Steady-State Characteristics

In Chap. 7 it was shown that the ratio of the torque developed at a slip s to the maximum developed torque is given by

$$\frac{T_d}{T_{dm}} = \frac{2s_p s}{s_p^2 + s^2} \tag{9-17}$$

This is simply a restatement of Eq. (7-30), where the slip at maximum developed torque is given by

$$s_p = \frac{R_2'}{X_2' + X_o}$$

and the maximum developed torque is given by

$$T_{dm} = \frac{qV_o^2}{2\omega_s(X_2' + X_o)} \tag{9-18}$$

where V_o is the Thevenin equivalent voltage, and X_o is the equivalent impedance since the primary resistance is neglected. When the slip is small the s^2 term in the denominator is negligible and the developed torque becomes

$$T_d = \frac{2T_{dm}}{s_p} s$$

$$= \frac{2T_{dm}}{s_p} \frac{(\omega_s - \omega_m)}{\omega_s}$$

and

$$\omega_m = \omega_s - \frac{s_p \omega_s}{2T_{dm}} T_d \tag{9-19}$$

If we substitute for T_{dm} in this expression and interpret it similarly to Eq. (9-4), we obtain

$$\omega_m = \omega_s - \left(\frac{\omega_s^2 R_2'}{qV_o^2}\right) T_d \tag{9-20}$$

$$= \omega_s - \frac{1}{B_o} T_d \tag{9-21}$$

where

$$B_o = \frac{qV_o^2}{\omega_s^2 R_2'} \tag{9-22}$$

The damping B_o therefore derives directly from the rotor resistance. Referring back to Chap. 7, it will be noted that Eq. (7-18), which gives the Thevenin voltage, becomes

$$V_o = \frac{X_m}{X_m + X_1} V_t$$

when the primary resistance is negligible, so that Eq. (9-22) may be rewritten to give the equivalent damping directly in terms of the primary phase voltage.

$$B_o = \frac{q X_m^2 V_t^2}{\omega_s^2 (X_1 + X_m)^2 R_2'} \tag{9-23}$$

When the subscript for the open circuit speed (ω_s in place of ω_o) is altered to suit Eq. (9-21), the circuit shown in Fig. 9-2 has this equilibrium equation. It is therefore a circuit model for steady-state operation of an induction motor.

Example 9-2

A three-phase, 220-V, 60-Hz, four-pole, wye-connected induction motor has the following parameters per phase:

$$R_1 = 0.44 \ \Omega$$

$$R_2' = 0.708 \ \Omega$$

$$X_1 = X_2' = 0.837 \ \Omega$$

$$X_m = 25.2 \ \Omega$$

Determine the operating speed when it is driving a mechanical system having a constant torque component of 7 N·m and a viscous friction component of 0.04 N·m·s and is connected to a 220-V source.

Solution. In the case of an induction motor, the value of the velocity source in Fig. 9-2 is always the synchronous speed, which is 188.5 rad/s in this case. The phase voltage is 127 V, so the Thevenin voltage is

$$V_o = \frac{25.2 \times 127}{(25.2 + 0.837)} = 122.9 \ \text{V}$$

and the value of the equivalent damping from Eq. (9-22) is

$$B_o = \frac{3 \times 122.9^2}{188.5^2 \times 0.708} = 1.801 \ \text{N·m·s}$$

From this point on, the solution is the same as that for the dc motor in Example 9-1. The operating speed is again obtained using superposition. The response due to the velocity source is obtained using the ordinary relation for a voltage divider given by Eq. (9-12). That is,

$$\omega_{mo} = \frac{1.801}{1.801 + 0.04} \times 188.5 = 184.4 \ \text{rad/s}$$

The change in operating speed due to the constant torque load is

$$\omega_{mT} = -\frac{7}{(1.801 + 0.04)} = -3.9 \ \text{rad/s}$$

The operating speed is therefore the algebraic sum of these two components:

$$\omega_m = \omega_{mo} + \omega_{mT} = 184.4 - 3.9 = 180.5 \text{ rad/s or } 1723.6 \text{ r/min}$$

9.3.2 Dynamic Characteristics

While the equivalent damping may be found from the steady-state circuit model, it is necessary to use the more general dynamic circuit equations to obtain the equivalent spring stiffness of Fig. 9-1, which again turns out to be the other parameter in an electromechanical model. The details are given in reference 6, where it is shown that this spring derives from the rotor leakage inductance.

$$K_o = \frac{q X_m^2 V_t^2}{\omega_s^2 (X_1 + X_m)^2 L_2'} \tag{9-24}$$

where L_2' is the secondary leakage inductance, referred to the primary.

Because the primary resistance has been neglected, the natural frequencies or poles associated with the stator transients are not included. As a result, any dynamic response involving significant stator transients is not predicted correctly by this model. In practice, this difficulty is limited to changes in terminal voltage at constant frequency where stator transients are bound to be present. In contrast, the transfer function relating a change in shaft speed to a change in load torque is predicted with little error. However, because the model is restricted to small values of slip, its application is restricted to changes in speed resulting from small perturbations in load torque, for example.

PROBLEMS

9-1. A 500-V, dc shunt motor has the following parameters:

$$M = 1.04 \text{ H}$$
$$R_a = 0.007 \ \Omega$$
$$R_f = 50 \ \Omega$$

Determine the operating speed when its armature and field are connected directly to a 500-V source and the load is
(a) a constant torque, $T_L = 16\ 000$ N·m,
(b) a torque proportional to speed, $B_L = 150$ N·m·s,
(c) a combination of $T_L = 10\ 000$ N·m and $B_L = 50$ N·m·s.
The effect of armature reaction may be neglected.

9-2. A 550-V, dc shunt motor has the following parameters

$$M = 1.37 \text{ H}$$
$$R_a = 0.005 \ \Omega$$
$$R_f = 220 \ \Omega$$

Determine the operating speed when its armature and field are connected directly to a 550-V source and the load is
(a) a constant torque, $T_L = 30\ 000$ N·m,

(b) a torque proportional to speed, B_L = 200 N·m·s,
(c) a combination of T_L = 15 000 N·m and B_L = 100 N·m·s.
The effect of armature reaction may be neglected.

9-3. A three-phase, 220-V, four-pole, 60-Hz, wye-connected induction motor has
the following parameters in ohms per phase

$$R_2' = 0.25$$

$$X_1 = X_2' = 0.4$$

$$X_m = 40$$

Determine the approximate operating speed when the motor is connected to
a 220-V, 60-Hz source and the load consists of
(a) a torque proportional to speed, B_L = 0.05 N·m·s,
(b) a constant torque, T_L = 10 N·m,
(c) a combination of B_L = 0.04 N·m·s and T_L = 2.0 N·m.

9-4. A three-phase, 460-V, four-pole, 60-Hz, wye-connected induction motor has
the following parameters in ohms per phase:

$$R_2' = 0.47$$

$$X_1 = X_2' = 0.63$$

$$X_m = 65$$

Determine the approximate operating speed when the motor is connected to
a 460-V, 60-Hz source and the load consists of
(a) a torque proportional to speed, B_L = 0.15 N·m·s,
(b) a constant torque, T_L = 30 N·m,
(c) a combination of B_L = 0.12 N·m·s and T_L = 10.0 N·m.

9-5. A 220-V dc shunt motor has R_a = 1.0 Ω, L_a = 0.01 H, and M = 0.8 H.
When connected to a 220-V source, its field current has a constant value of
1.5 A. The moment of inertia of its armature and load is 0.2 kg·m².
(a) Assuming that the armature current on no-load has a negligible effect,
determine the no-load speed of the motor.
(b) A load torque of 18 N·m is suddenly applied. Determine the system
function relating the change in speed to the load torque.
(c) Calculate the time constants of the resulting response.
(d) Use the system function and the answer to part (a) to determine the final
speed of the motor.

9-6. If the motor of problem 9-5 is connected to the load by a shaft that is not
rigid but has a stiffness of 1000 N·m/rad, determine the transfer function
relating the change in speed to the load torque. Assume that the inertia
is equally divided between the motor armature and the load, namely
0.1 kg·m².

9-7. A 500-V dc shunt motor has R_a = 2.0 Ω, L_a = 0.02 H, and M = 1.5 H.
When connected to a 500-V source, its field current has a constant value of
2.5 A. The moment of inertia of its armature and load is 1.3 kg·m².
(a) Assuming that the armature current on no-load has a negligible effect,
determine the no-load speed of the motor.
(b) A load torque of 30 N·m is suddenly applied. Determine the system
function relating the change in speed to the load torque.

(c) Calculate the time constants of the resulting response.

(d) Use the system function and the answer to part (a) to determine the final speed of the motor.

9-8. A three-phase, 460-V, 60-Hz, four-pole, wye-connected, wound rotor induction motor has the following parameter values per phase.

$$R_1 = 0.5 \ \Omega \qquad X_1 = 1.5 \ \Omega$$
$$R_2' = 0.7 \ \Omega \qquad X_2' = 1.5 \ \Omega$$
$$X_m = 40 \ \Omega$$

The moment of inertia is 5.0 kg·m². It is driving a constant torque load at a slip of 0.02 when the load torque is suddenly increased by 3.0 N·m. Determine the transfer function relating the change in speed to the change in torque, and hence the time constants of the resulting transient. Obtain the final value of the change in speed.

9-9. A three-phase, 550-V, 60-Hz, six-pole, wye-connected, wound rotor induction motor has the following parameter values per phase.

$$R_1 = 0.9 \ \Omega \qquad X_1 = 2.5 \ \Omega$$
$$R_2' = 1.2 \ \Omega \qquad X_2' = 2.5 \ \Omega$$
$$X_m = 55 \ \Omega$$

The moment of inertia is 6.0 kg·m². It is driving a constant torque load at a slip of 0.03 when the load torque is suddenly increased by 5.0 N·m. Determine the transfer function relating the change in speed to the change in torque, and hence the time constants of the resulting transient. Obtain the final value of the change in speed.

9-10. If the motor of problem 9-9 is connected to the load by a shaft that is not rigid but has a stiffness of 1000 N·m/rad, determine the transfer function relating the change in speed to the load torque. Assume that the inertia is equally divided between the motor armature and the load, namely 3.0 kg·m².

REFERENCES

1. Kron, G. *Tensors for Circuits.* New York: Dover Publications, Inc., 1959.

2. White, D. C., and H. H. Woodson. *Electromechanical Energy Conversion.* New York: Wiley, 1959.

3. Lindsay, James F., and Silas Katz. *Dynamics of Physical Circuits and Systems.* Portland, Oregon: Matrix Publishers Inc., 1978.

4. Lindsay, J. F. "An Electromechanical Network Model of the DC Motor," *IEEE Transactions on Industry Applications,* IA-14, no. 3 (May/June 1978): 227–33.

5. Rashid, M. H. "Dynamic Responses of DC Chopper Controlled Series Motor," *IEEE Transactions on Industrial Electronics and Control Instrumentation,* IECI-28, no. 4 (November 1981): 323–30.

6. Lindsay, J. F., and K. Venkatesan. "An Electromechanical Network Model for Frequency Controlled Induction Motors," *Electric Machines and Electromechanics* (June 1981): 225–38.

BIBLIOGRAPHY

CHAPMAN, C. R. *Electromechanical Energy Conversion*. New York: Blaisdell Publishing Co., 1965.

CHAPMAN, S. J. *Electric Machinery Fundamentals*. New York: McGraw-Hill Book Co., 1985.

DEL TORO, V. *Electric Machinery and Power Systems*. Englewood Cliffs, N.J.: Prentice-Hall, Inc., 1985.

ELLISON, A. J. *Electromechanical Energy Conversion*. London: Harrap & Co. Ltd., 1965.

FITZGERALD, A. E., C. KINGSLEY, and S. D. UMANS. *Electric Machinery*. 4th ed. New York: McGraw-Hill Book Co., 1983.

GOURISHANKAR, V. and D. H. KELLY. *Electromechanical Energy Conversion*. 2nd ed. New York: Intext Educational Publishers, 1973.

HINDMARSH, J. *Electrical Machines and Their Applications*. 3rd ed. New York: Pergamon Press, 1977.

MABLEKOS, VAN E. *Electric Machine Theory for Power Engineers*. New York: Harper & Row, Publishers, Inc., 1980.

MAJMUDAR, H. *Electromechanical Energy Converters*. Boston: Allyn and Bacon, Inc., 1965.

MATSCH, L. W. *Electromagnetic and Electromechanical Machines*. 2nd ed. New York: Harper and Row, Publishers, Inc., 1977.

MEISEL, J. *Principles of Electromechanical Energy Conversion*. New York: McGraw-Hill Book Co., 1966.

NASAR, S. A., and L. E. UNNEWEHR. *Electromechanics and Electric Machines*. New York: John Wiley & Sons, Inc., 1979.

SCHMITZ, N. L., and D. W. NOVOTNY. *Introductory Electromechanics*. New York: Ronald Press Co., 1965.

SKILLING, H. H. *Electromechanics*. New York: John Wiley & Sons, 1962.

SLEMON, G. R., and A. STRAUGHEN. *Electric Machines*. Reading, Mass.: Addison-Wesley Publishing Co., Inc., 1980.

SMITH, RICHARD T. *Analysis of Electrical Machines*. New York: Pergamon Press, 1982.

THALER, G. J., and M. L. WILCOX. *Electric Machines: Dynamics and Steady State*. New York: John Wiley & Sons, Inc., 1966.

INDEX